Y0-BKP-105

Health at Work

Health at Work

WARD GARDNER and
PETER TAYLOR

A HALSTED PRESS BOOK
JOHN WILEY & SONS
New York — Toronto

English language edition except USA and Canada, published by
Associated Business Programmes Ltd
17 Buckingham Gate, London SW1

ISBN 0–470–29144–3

Published in the USA and Canada by
Halsted Press, a Division of
John Wiley & Sons Inc
New York

First published 1975

Gardner, Archibald Ward.
    Health at work.

    "A Halsted Press book."
    Includes bibliographies.
    1, Industrial hygiene. 2, Industrial safety.
I. Taylor, Peter, joint author. II. Title.
(DNLM: 1. Occupational diseases – Popular works. WA400 G226h)
RC967.G37   613.6'2   75–5979

© Ward Gardner & Peter Taylor

This book has been printed in Great Britain
by Flarepath Printers Ltd, St Albans, Herts,
and bound by W. J. Rawlinson Ltd, of London

Contents

PREFACE	VII
ACKNOWLEDGEMENTS	IX
1. WHY HEALTH IS OF CONCERN TO MANAGERS AND WORKPEOPLE	7
2. HOW INTEREST IN HEALTH AT WORK BEGAN TO GROW	13
3. THE ORGANISATION OF HEALTH SERVICES AT WORK	18
4. ROUTINE EXAMINATIONS AND SCREENING TESTS	25
5. ABSENCE: SICKNESS OR MALINGERING?	33
6. DISEASE, IMPAIRMENT AND DISABILITY	43
7. SHIFT WORK AND HEALTH	53
8. OCCUPATIONAL HAZARDS AND HOW TO DEAL WITH THEM	62
9. CHEMICAL HAZARDS	78
10. PHYSICAL HAZARDS	98
11. SAFETY AT WORK—PREVENTING INJURIES AND DAMAGE	113
12. THE WORKING ENVIRONMENT	127
13. MENTAL HEALTH AND MOTIVATION AT WORK	134
14. SOME SPECIFIC OCCUPATIONAL HEALTH PROBLEMS	143
Mortality rates, executives, drivers, food handlers, backache, eye hazards, infections, hospital staff, drugs, alcohol.	
APPENDIX I Some Journals on Occupational Health published in English.	159
APPENDIX II Draft Form for Investigating and Reporting Incidents.	161
INDEX	164

To Elaine and Jo, who have suffered much
in the cause of Occupational Health.

Preface

This book is written for non-specialists to provide an overall view of health and a brief scan of safety in business and industry. We have included thoughts about both, because safety and health are inexplicably intertwined and can only be viewed properly as a whole. There is no attempt at exhaustive covering of work hazards or occupational disease – this would only fatigue the reader's interest. Instead, we have tried to indicate the kinds of approach to problems of health and safety which we believe to be appropriate, to make general points by example, and to give perspective to health and safety at work. We believe that the views expressed in this book are not far from the mainstream of modern thought on most topics. This does not, however, rule out occasional idiosyncrasies. Throughout the book we have used the word 'manager' in the sense of the person who is in charge and who is responsible for the work of others: in this sense 'manager' includes chairmen and managing directors, managers, supervisors, foremen and others in positions of authority.

A list of books for further reading appears at the end of each chapter. Authoritative references and scientific evidence about the topics covered in the chapter may be found in these books by those who may wish to delve more deeply into the subject. We have not, however, in a book of this size intended to do more than signpost useful references. No attempt is made to cover any subject in depth. Also added at the end of the book list is a list of some of the more useful journals on occupational health published in English.

In order to be useful a text should cover the appropriate ground in a balanced way, and must be accurate and up-to-date. We would greatly appreciate comment and criticism from readers about the content and coverage of this book, and would be glad to have errors called to our attention. Please write to us if you have any comments c/o Associated Business Programmes, 17 Buckingham Gate, London SW1.

We would like to emphasise that the views and opinions expressed in this book are our personal ones and should not be taken to represent the views or opinions of any organisation with which we are, or have been, associated.

Ward Gardner and
Peter Taylor
London, 1975

Acknowledgements

We wish to thank the following people for their invaluable help in the preparation of this book: Reg Beasley, Joyce Cooper, Brian Dagnall, Dorothy D'Ardenne, Phyllis Garment, Cathy Girvin, Jim Sanderson, Marion Self, John Thackway, Eve Williams and Alex Young.

CHAPTER 1

Why Health is of Concern to Managers and Workpeople

Occupational health/safety problems at work today

Most of the illnesses and injuries which arise daily from work result from known and well-recognised hazards for which suitable countermeasures have been worked out. In other words, most of the illnesses and injuries are not related to new or unrecognised phenomena. They may be new to the people concerned or they may have been unrecognised by those in charge. The managers' problem, therefore, is one of awareness and recognition coupled with the need to get expert help and guidance if he suspects a problem. Similarly, in the case of new problems, he should always look for the worst possibility. It is, for example, unwise to assume that any substance is non-toxic until it is proved to be so. Many substances can give rise to allergies, skin trouble, or if volatile can lead to problems if inhaled. All of these risks beg the question 'who is going to be in contact with the substance and how is it going to be used?'

Apparently simple changes in a situation can produce very large changes in hazards. For example, a furniture workshop changed from animal glue to epoxy glue. Within a short time, about half of the workforce had developed allergic dermatitis from the epoxy glue so that they could no longer be employed. Animal glues can safely be slapped about and left on the skin: epoxy glues can produce severe allergic skin manifestations if allowed to get on to the skin and to remain there.

Or in a printing works, a young worker found that benzene was a good solvent for cleaning tacky material off rollers and ordered some for this purpose. In a smallish workroom, this produced quite high concentrations of benzene vapour and he breathed this atmosphere regularly. Inside of a year he was in hospital with severe anaemia and until the type of anaemia was diagnosed, no one thought that the illness could be due to his work.

A third example occurred in a part of a factory drawing wire and sending out lacquered wire. A problem arose because a number of men developed wheezing and asthmatic-type symptoms in the work area. The lacquer drying was done by heating, the exhaust of which was vented into the top of the work space, not through the roof. Investigation revealed that the type of lacquer used had been changed recently and the new lacquer contained diisocyanates which were driven off by heating and were vented into the general air of the workroom. Extremely small amounts of diisocyanates are capable of producing spasm of the small bronchial tubes and thus lead to complaints of wheezing, chest tightness and asthmatic attacks. Putting the exhaust from the driers through the roof prevented further medical problems. The understanding of why the symptoms arose and how these could be prevented avoided the strike threat which was developing amongst those employed.

The trend of legislation and code of practice

Legislation, codes of practice and the like are everywhere becoming more demanding in the fields of health and safety. It is therefore important that managers keep up with, or preferably ahead of, minimal standards in order to remain in business. There are, of course, other important reasons for interest in safety, health and ecology which can be stated under the heading of normal responsibilities both to workpeople and to the public at large.

People at work

Today, people are not regarded as expendable, though in some parts of the world and in certain bad conditions in so-called civilised communities examples of exploitation can still be found. In principle, however, no right-minded manager will behave in that way, and no reputable organisation will tolerate such views. People should not be made ill, or be injured by work. Work environments should be safe and healthy. Toxic substances, for example, should only be handled in ways that are known to be safe. The human relationships within any organisation should be fair and equitable, with a minimum of imposed decisions from above and of unrealistic group attitudes from below.

All of these things can contribute to overall mental and physical health or ill-health. Managers can influence and direct much of what happens in these areas. People at work are like people anywhere—they respond to the same things in positive and negative ways and they are favourably or adversely influenced by their surroundings. While these statements may appear platitudinous and self-evident, we should still like to emphasise the need for recognition by managers of adverse or unfavourable working conditions. With recognition of such problems, action to remedy the situation can follow; without, the situation continues.

Many people today are, for example, working in noise where the overall level is more than 90 dBA. They may be working as drivers of vehicles or working in manufacturing or in service industries. The likelihood of noise-induced deafness increases as the noise level rises. Today we know how to measure noise levels, how to relate noise measurement to predicted hearing damage and how to predict safe durations of exposure. We also know how to design plant and equipment to reduce noise generation at source and how to protect individuals against damaging noise by reducing noise exposure.

Unfortunately, ship manufacturers are still turning out ships which will deafen those who work in the engine room and designers of industrial plant or equipment seldom provide noise specifications. At present only a minority think of noise as a problem to be considered at the design stage. In many cases the people who work in ships and who use the equipment are quite unaware of the risk to their hearing. Have management a responsibility here? Should legislation be introduced to ensure that safe working conditions are provided? And if safe conditions cannot be provided should the wearing of ear muffs or other hearing protection be mandatory? These questions raise many other problems. What should the attitude of management organisations and trades unions be towards health and safety? We have seldom met an industrial or trade union leader who, in speaking about the subject, was not convinced of the need for safe and healthy working. This is splendid. But we do not usually find a large cross sectional group of the whole of management or of union and shop-floor representatives who are equally committed. There are, of course, some fine men and committed groups who stand out by their concern. In general, however, the commitment to safety and health often seems to get rather lost or diluted. Perhaps top managers and leaders of trade unions may see this as a task which requires their attention.

Why health is of concern to managers and workpeople

In the final analysis productivity, organisational viability and profit depend on people and how well or how badly they perform. Performance will be related to a number of factors, one of the most important of which will be health—both physical and emotional. In terms of general health, managers and workpeople have the same responsibilities in looking after their own health. Is this responsibility always adequately discharged by individuals? Numerous examples of obesity from overeating, of chronic bronchitis, lung cancer and heart disease resulting from cigarette smoking and so on could be given. Such individuals will, of course, tend to be less effective because of more time off due to sickness. A recent article in a medical journal commented on the fact that an average of nearly one-third of the patients in an acute medical ward were there by virtue of their living habits in matters such as inadequate or poor diet, cigarette smoking disease, obesity, alcohol-related illness and so on.

The information about how to live healthily is not yet applied or understood by large sections of the population. How much responsibility any manager assumes for the personal health problems of his subordinates, or how much it is the business of any organisation to concern itself with the personal health of its employees is a matter for debate. There are some organisations which show no evidence of concern, while others run elaborate schemes of periodic medical examinations for some or all of their employees. Around this has grown the mythology of executive health or ill-health. Some managers, and some workpeople, especially those with union responsibilities, can become very excited and concerned when they find any work hazard which, even in a slight way, could make a person ill or injure him. We are, of course, in agreement with this concern, but we sometimes find that those same people are obese, smoke large numbers of cigarettes and drive without wearing their seat belts. Their interest in health and safety would be more plausible if they practised what they preached. It could also benefit them! Example is still a potent influence.

The manager's responsibility for health

In any business or industry management are responsible for the creation of the working environment, both physical and emotional. The qualities —positive or negative—of these environments can preserve, enhance or adversely affect the health of employees and others both inside and outside of the immediate area. Public awareness of pollution, product hazards and noise, to give only a few examples, is increasing all the time. With this greater public awareness comes an intolerance of unnecessary dangers both personal and environmental.

The table below shows some of the responsibilities for health which may have to be shouldered by managers.

These responsibilities are not always seen or acted upon by directors, managers, supervisors or companies as industrial diseases, injury at work and environmental pollution continue to occur. Omission of precautions to prevent trouble in any of these areas may lead to unfortunate consequences. Any view of the management responsibility for health must, therefore, look beyond both the immediate physical health of employees and the problems of the working environment.

Recent examples of noise arising in factories becoming the cause of litigation by neighbourhood residents, of downstream pollution of waterways by industrial effluents, of unsafe substances inadequately tested

Why Health is of Concern to Managers and Workpeople

FIGURE 1: *Manager's responsibilities for health*

Whose health	Mental health	Physical health	Work environment factory, office, plant
Directors, managers and supervisors	yes	yes	yes
Employees	yes	yes	yes
Contractors	—	yes	yes
Visitors	—	yes	yes
The public	—	possibly	possibly if any access

and labelled being supplied to the public are enough to make many people aware that managers have significant responsibilities in these matters. They also realise that if these responsibilities are evaded or omitted, legal remedies are available. The forces of public opinion are now being brought to bear with considerable force on these who are seen as offenders.

Doctors, occupational hygienists, nurses, design engineers, and many others can advise and help managers in the discharge of their responsibilities for health and safety, but the final decision in these matters must remain with the manager.

Why is health of concern to workpeople

Every individual has a responsibility for his own health which he cannot pass to others. To remain healthy, he must follow the basic rules of health which include eating properly, getting enough sleep and exercise, not smoking, limiting the amount of alcohol taken and generally living in a way which minimises stress and strain and allows time for relaxation and hobbies. At work the rules for health will include ensuring as far as possible that hazards are identified and precautions taken, both for himself and for his colleagues.

Unfortunately, many people both at work and away from work do not follow these basic rules. As a result, some suffer. It may appear self-evident that people would always act in their own interests in matters of health and safety. Unfortunately, this is not so as a review of risk-taking behaviour as seen, for example, in driving or in relation to safety and health rules will soon show. However, the reasons why health *should* be of concern to all workpeople are not less valid because some people disregard the rules. They, or others, may easily suffer as a direct result of this lack of concern.

The functions of an occupational health service

The International Labour Organisation (ILO) and the World Health Organisation (WHO) set out in 1959 what they believed to be the functions of an occupational health service. The first statement under this heading is 'the role of occupational health services should be essentially preventive'. With this we would agree and would also wish to underline its importance. A brief list of basic functions may be useful at this stage to outline the main areas in which occupational health services should operate:

1. To identify occupational hazards.
2. To advise on the control of occupational hazards.
3. To recognise at an early stage occupational or other disease and to screen vulnerable groups if advisable.
4. To give initial treatment for injuries and illnesses of sudden onset; to treat illnesses to prevent people from going off work. (It is not the role of an occupational health service to become deeply involved in routine medical treatment.)
5. To give advice about the placement of people in suitable work—on starting work, following illness or injury, or at other times when problems may exist.
6. To provide general advice and supervision of conditions at work which may influence health, such as food handling and eating facilities and general sanitation.
7. To undertake health education.

Most of these functions—and especially the preventive ones—are dealt with in detail later. It is, however, important to realise now that occupational health services can only be a part of a team effort to create and maintain healthy work places with healthy people. Design engineers, chemists, architects and so on have at least as much to do with health at work in many situations as do people who are recognised as members of a medical group. Failure to communicate and/or co-operate adequately is often a source of problems.

Further reading

A Management Guide for Occupational Health Programs, American Medical Association, Council on Occupational Health (1964).

Advantages of a Company Health Service, British Employers Confederation, London (1964).

Occupational Health: a guide to sources of information, Suzette Gauvain, Society of Occupational Medicine, London (1968). Wm. Heinemann Medical Books Ltd., London, 1974.

CHAPTER 2

How Interest in Health at Work Began to Grow

It is surprising that Hippocrates, who based his theory of medicine upon the interactions between man and his environment, should have made no mention of work. He viewed sickness as the result of an imbalance between the environmental 'humours' of air, earth, water and fire, with the bodily humours of blood, phlegm, yellow bile and black bile. The reason for his omission of work was probably the attitude of the Greeks towards work which was that it should only be done by slaves. It is probably because of this that concern by doctors for the health of workers did not start to become mentioned in medical texts until the Middle Ages when two medical books were published which included sections on the diseases of miners.

There is an interesting historical link between the book on metals by Agricola (1556) and many modern currencies, and the hazards of radiation. He described the mining of silver in the Joachimsthal region of Bohemia and mentioned the lung disease that many of the miners contracted. The name of the area was used to describe the Austrian silver coin 'thaler' which became corrupted for international use as 'dollar'. The area was subsequently found also to produce a radioactive ore and some, at least, of the miners' lung diseases was due to radiation-induced lung cancer.

Occupational medicine accepts Bernadino Ramazzini as its 'father', since in 1801 he wrote what was the first textbook on the subject. Although he attributed some diseases to the offensive smell of certain jobs, he described the action of many of the commoner toxic substances such as lead and mercury, and made many recommendations for their control. He also appreciated the need for what is now termed ergonomics, by suggesting modification of equipment to better suit the capacity of men. His greatest contribution to medicine was to suggest that when examining patients, doctors should add to the questions that Hippocrates had laid down the critical one 'what is your occupation?'

It is perhaps inevitable that most of the better-known industrial diseases should have become common during the industrial revolution when large numbers of workpeople were gathered in factories in which neither the owner nor the law appreciated the nature of the hazards. However, many of the more serious effects of this change in society were not strictly occupational in origin. The family life and reasonable nutritional standards of the rural seventeenth and eighteenth centuries were disrupted when the men moved into the new factory towns. Overcrowding, a total lack of sanitary engineering, poor-quality food and the long hours of work between them produced conditions ideal for the spread of infectious disease. Cholera and smallpox were rife in most towns; typhoid, typhus and tuberculosis, together with the less dramatic infectious diseases, combined to reduce the expectation of life of adults and raised the infant mortality rate to unprecedented levels.

The harnessing of steam power to move machinery increased the speed of manufacturing and work injuries became commonplace. Little, if any, attention was paid to safety. It took many years to persuade or force employers to guard, for example, the driving belts which could and did whisk a person to the ceiling and crush him between belt and wheel.

Long hours of work—16 hours a day or 100 hours a week—were commonplace. The inevitable consequences of malnutrition and fatigue included high sickness and injury rates. Thackrah, a doctor who interested himself in diseases arising from work, noted that the knife grinders in factories in Sheffield seldom lived beyond the age of thirty. They worked in rows lying down with their heads close to the sandstone grinding wheels where they inhaled large quantities of freshly fractured sand particles. This caused silicosis which, combined with tuberculosis, managed to kill them very quickly.

It is ironic that the advent of the First World War with its large-scale slaughter probably did more to improve standards of occupational health and safety in the factories of the combatant countries than many years of patient legislation. The desperate shortage of manpower and the need for maximum output of munitions led governments in all the countries involved to commission research into finding solutions. The report in 1917 of the Health of Munitions Workers Committee in the United Kingdom, for example, demonstrated quite clearly that the traditional two hours of work before breakfast was inefficient and that production fell and accidents rose when more than sixty hours were worked each week. The hazards to health from dust and certain chemicals were clearly identified and a great deal of basic research was completed on industrial accident and injury patterns.

The Second World War, too, produced substantial developments in occupational health and safety. Ergonomics, the science concerned with adapting machines and their controls to the mental and physical attributes of man, dates largely from this time. Studies of this nature are made at times when manpower is short and when the best use must be made of limited resources. It is a sad reflection on society that concern for the health and safety of people at work is greatest when there are military or economic reasons for taking this view. Many of the 'third world' countries are now in the stage of industrial revolution, and for them the problems of urbanisation and the disruption of a rural and family-orientated society are being compressed into a few decades, rather than the century or more during which these changes occurred in the countries of Western Europe and the USA. Many of the under-developed countries have large resources of unskilled labour and are having to deal with exploitation of workpeople by entrepreneurs having a scant regard for the health and safety of their workpeople.

The idea of working at regular times and of following fairly inflexible routines of occupation and leisure times is of comparatively recent origin, dating mainly from the Industrial Revolution. Before this, while a few people had to work to a schedule, most worked only when there was the need to do so. When the job was finished, they stopped working. The time pressures generated were simply those of getting the work done. The rate of work was largely self-determined. With the advent of the Industrial Revolution, all this changed for those people who had to tend machinery. Workpeople were seen as servants of the technology and had to adapt to it. This idea of the person accommodating to the job overflowed into many work situations in which no machinery was involved, for example, office hours were rigidly fixed for work, some of which could be done as well at midnight as midday if this had been allowed. Flexible working hours, giving people some choice of working hours around a central core of required attendance, is hailed as a modern concept; in fact it is largely a reversion to a more ancient method of working in which the human being is not bent completely to the system. Self-employed people and senior managers have, of course, worked flexible hours all through this period.

The modern notion is of a socio-technical system in which the strengths and weaknesses of people as individuals and in groups are correlated with the various facets of the technology in such a way as to make the best use of both. The idea of paying attention to the interface between technology and people and of recognising the importance of the sociological considerations arising from such a socio-technical system, were slow to emerge as important concepts for industrial work. Indeed, these ideas still find scant or little recognition in some places today. The grouping of large numbers of people at work also tends to date from the Industrial Revolution. The potential hazards of unfavourable environments, harsh conditions of work, fatigue and malnutrition were greatly enhanced among those who were herded into the 'dark satanic mills' described by Blake. Many other work places such as 'sweat shops' were equally grim both in their physical environment and in the ways that the human beings who worked there were treated (the psycho-social environment).

Doctors were first invited into factories because a few of the more humanitarian employers became concerned about the diseases and injuries occurring in their workforce. The diseases were not those usually strictly associated with occupation but were more often those associated with malnutrition, poor living conditions and non-existent sanitary hygiene. Death rates were high from such conditions as rickets, tuberculosis and the enteric diseases. Robert Owen engaged a doctor and a clergyman to look after the 'bodies' and 'souls' of his employees in his New Lanark Mills at the end of the eighteenth century, but he and the Quaker employers were exceptional. Outside pressures, too, began to gather about the conditions under which people were forced to labour. The interest of these early doctors, therefore, centred on *diseases*, whether occupational disease like lead, mercury or phosphorus poisoning, or diseases which were related to occupation but which could be described more accurately as community diseases from which certain working populations suffered a higher incidence due to the conditions under which they lived and laboured.

Charles Turner Thackrah (1795–1833), who worked in Leeds, devoted his short life to the study of occupational disease and the hazards which arose from industrialisation. He placed much emphasis on prevention saying, 'Each master . . . has in great measure the health and happiness of his workpeople in his power . . . let benevolence be directed to the *prevention*, rather than to the relief of evils, which our civic state so widely and deeply produces.' His systematic observations on industrial disease and its prevention, which he wrote up in a book *The Effects of the Principal Arts, Trades and Professions and of Civic States and Habits of Living on Health and Longevity* (1831), played an important part in stimulating people who were interested in social reform. In the long term his work led to factory and health legislation and to some of the social reforms of the late nineteenth century.

As the social consciences of people and nations awoke to the ills arising from factory work, legislation concerning standards of hygiene at work, hours of work, control of occupational hazards and so on was enacted in many countries. This legislation usually appointed inspectors: first to see that the law was not broken and at a later stage to advise on health problems and to supervise hazard control. In Britain, for example, doctors first became legally involved in factories in order to certify that any children employed were of the physique and appearance of a child of nine years. It took another fifty years for doctors to become legally required to investigate and advise on prevention of specific diseases. It is only in the past few years that they have had their legal role extended.

Occupational health is served by two main disciplines: occupational medicine and occupational hygiene. Occupational medicine deals mainly with *people* and with the two-way interaction between work and health. Occupational hygiene—or environmental engineering as it is called in some countries—is concerned mainly with the physical and chemical *environment* at work. Hygienists are trained to measure, evaluate and advise on the control of hazards in workplaces. These two disciplines are complementary and with the inclusion of safety and accident prevention complete the team for the prevention of injury or ill health at work.

Further reading

Occupational Health Practice, R. S. F. Schilling (ed.), Butterworth & Co., London (1973).

The Diseases of Occupations, 4th edition, Donald Hunter, English Universities Press, London (1969).

De Morbis Artificum, B. Ramazzini (1713), translated by W. C. Wright, University Press, Chicago (1940).

The Effects of Arts, Trades and Professions and of Civic States and Habits of Living on Health and Longevity, C. T. Thackrah, 2nd edition, Longmans, London (1831).

The Hazards of Work: how to fight them, P. Kinnersley, Pluto Press, London (1974).

CHAPTER 3

The Organisation of Health Services at work

Health services at work

Health services at work will vary according to size of the work group, the hazards involved, the location of the plant or process in relation to any community medical services, and many other factors. In any individual case, the important thing is to get the right amount of medical service of the correct kind. For example, a factory engaged two nurses who were under-employed to the point of boredom, being expected to provide a first-aid and casualty service only. They were not allowed to go round the work areas unless sent for to attend to a casualty. In another factory, however, with very dangerous hazards requiring close medical supervision of workpeople, the management had taken advice from a doctor who was expert in the field of occupational health. As a result, a doctor was engaged—not the nearest doctor, or the one who looked after the manager's wife and family but one who was trained and experienced in occupational health—to provide the two hours per day of his time which was all that was needed. This doctor was given further specialised training about the particular hazards in that factory, and was able to call on the expertise of the doctor first called in by the management to advise on the problem. In this way, a preventive occupational service was set up tailored to the needs of the factory and its particular hazards and making best use of adequately trained people.

Introducing occupational health services

The first step in this, as in many other problems, is to define the problem and to assemble information which can be used to decide the type of service best suited to local needs. Expert help at this stage can save a lot of time, money and subsequent problems. Management should have some ideas about the type of service which they want—and for which they will have to pay. Do management want to provide only the minimal basic services necessary, or would they see the provision of health services at work to be a way of helping to foster high standards of mental and physical health or a useful fringe benefit? The point to make in any discussion about such services is that grades of services can be provided, just as when buying a car the choice extends from lavish status symbols to out-of-date and barely efficient models. Decisions about the grade and quality of the service can only be made in a sensible way by joint discussion between management, and those who can advise in an expert way on the provision of services or those who actually provide them. Existing occupational health services can often benefit by review and answering questions such as 'what are we trying to do', 'why are we doing it this way', or 'could it be better done in other ways'.

Personnel for occupational health services

The brief review of the history of occupational health services at the beginning of the book showed that the interests of doctors working in industry were first in disease and later in health and prevention. In selecting doctors and nurses for work in occupational health today, special attention must be paid to training in preventive medical care and in occupational health. The care of healthy people in groups is a principal function of any occupational health service, and quite unlike the care of people as individuals when they are ill. It is for this reason that the selection of doctors and nurses from backgrounds of illness-orientated or curative medicine, such as general practice or hospitals, is not always successful unless either adequate supervision can be given on a day-to-day basis by suitably trained people, or further training can be given to those doctors and nurses in the techniques of the application of health care in groups. Neglect of this principle leads to needless lowering of the quality of service given and to curative instead of preventive activities being stressed in the overall programme. It is also important that those appointed should be able to understand or wish to learn enough of the technology of the industry to talk to the technical experts about the health problems which arise. The contributions to health which can be made by occupational hygienists, design engineers, sound and vibration experts, chemists and so on must be appreciated, because failure to use their skills will result in standards of performance below the best available.

In good occupational health services the emphasis is on prevention of ill health, whether arising from work or from other causes. The aim must always be to keep the people at work and to prevent them from becoming ill—on their feet rather than on their backs. Any medical treatment given at work should serve these aims. The treatment of long-term illness can be extremely expensive and since doctors at work cannot undertake care of workpeople in their homes, this is far better left where it properly belongs as the responsibility of the family doctor. Some senior managers appear to have chosen their doctor with the motive of obtaining private medical care for themselves and their families. This may be for them an excellent fringe benefit but, as many doctors and managers have found out from bitter experience, it is a sure recipe for a poor quality occupational health service. If in doubt a manager can seek advice from a national medical association or society of occupational health in his own country.

Occupational health services

The objectives and responsibilities of an occupational health service can only be as good or as bad as the company policy allows. So, before detailing any of these it is necessary to set out a statement of medical policy which could for example be, 'The company believes that the health of its employees is a matter of primary concern. Company policy is therefore to provide healthy and safe working conditions and to encourage each employee to maintain good health and to work safely.' Following on from this policy, the aims of an occupational medical service could be stated thus:

1. To protect and improve the physical and mental health of all workpeople and assist management to implement company policy.
2. To explain and discuss matters related to the interaction between work and health with individuals, management and unions.
3. To identify, assess and advise management on the control of any health hazards affecting employees or the public which may arise from the company's activities.
4. To advise on the effects of ill health on the working capacity of all staff at the stage of recruitment, during employment and on retirement.
5. To provide, within agreed limits, an individual service to each person in the organisation for initial and follow-up treatment of illness and injury.
6. To respect the confidentiality of personal medical information and act in the best ethical standards of the professions involved.

It is not necessary to be a large multinational company to have clear policy aims, nor is it necessary even to have full-time doctors or nurses in order to do the job properly. But the policy statement is an essential beginning to any service which aims to be of good standard because it allows everyone to understand what is being attempted.

In setting up occupational health services the person in charge of these services, whether a doctor or a nurse must have easy access to senior management on a confidential basis. In the case of a doctor (or in his absence, the nurse) it is probably best that he should be directly

The Organisation of Health Services at Work

responsible to the senior manager in the area in which he is working. If the organisation has a central occupational health group, a functional or 'dotted line' responsibility to the head of this group is also essential.

It is reasonable that managers and supervisors should expect guidance on a wide range of problems from their occupational health advisers. Yet, often, the expectations of managers, supervisors and people in general in organisations which do not have or have only recently acquired an occupational health service is too low. Certainly this is bad for the organisation and it is stultifying for good people not to have their skills used. Managers should, therefore, try to see that occupational medical and nursing staff have opportunities to present their views on a regular basis so that each can understand the problems of the other, and in this way the standard of expectation and usefulness can be raised.

A guide to the numbers of occupational health staff

The number of people required to produce an occupational health service in any given place will vary with the standard of service expected and with the hazards of the industry. Work situations which are relatively static, in which the hazards are well recognised and possibly not great, will obviously require a different coverage from a situation in which a wide range of potentially serious hazards are present and where problems are far from static.

One nurse can be appointed to deal with somewhere between 500–2000 workpeople. She may need, and will usually be glad to have, the help of a part-time doctor for advice and consultation for a few hours per week, or as required. A full-time doctor can be expected to deal with the occupational health problem of about 2000–7000 workpeople, depending on the help which he receives from nurses and other auxiliaries, and on the nature and hazards of the work. These guides must be seen as very broad generalisations: the best way to deal with any individual problem is to have expert assessment, taking into consideration all the known facts.

What should managers expect of their medical and nursing advisers?

The importance is again stressed of seeing occupational health services as group services with a strong bias towards prevention, and an overlap in function with other preventive activities such as design of workplaces, safety activities of all kinds and accident prevention as a whole. Too often the role of the doctor or the nurse is seen in terms of first-aid finger-bandaging or in treating illness which arises at work. While these are important activities and should be done well, it is vital that they should only be seen as part of the total activity. Managers should expect occupational medical and nursing staff to take a real interest in all the work which is done, about all the processes which are carried out and about all the people in the organisation. They must become part of the organisation, understand its management philosophy and be able to give advice based upon a real knowledge of the industry, factory or plant.

In any dealings with individuals, managers should recognise that medical confidences have to be respected. This should not, however, be a barrier to an intelligent exchange of mutual problems between medical and non-medical people who have reasonable and legitimate interests in these problems. Provided that both sides respect the need for individual medical confidentiality and are prepared to discuss matters in a problem solving way, solutions to most situations can easily emerge. Doctors and nurses in industry can discuss an environmental problem and its solution, or what a person could or could not do at work, without any mention of medical diagnosis or why and should not retreat behind unnecessary screens of confidentiality. Managers, for example, have a need to know how long a member of their staff is likely to remain off work if he is ill and if, when he returns, he is likely to be able to do his normal job.

Professional training in occupational health and hygiene

The general medical or nursing training contains very little in the way of specialist training in occupational health. The attitudes of many doctors and nurses are geared towards *treatment* of the *individual* who is sick. A change in objectives is required if they are to practice good occupational health which is primarily concerned with the *prevention* of ill-health in *groups* of people at work.

Specialised courses of training in occupational health for both doctors and nurses are available in most industrialised countries. These are usually arranged by academic institutions and recognition as a specialist in many countries requires that the doctor shall have had both formal training and adequate practical experience. In-service training can, of course, be most useful, but it usually fails to cover certain areas or hazards which do not occur in that particular organisation. We also believe that more emphasis could with benefit be given to an understanding of mental health and motivation at work (p. 134).

Managers should encourage their doctors and nurses to belong to their appropriate professional society and to attend meetings on a regular basis. Contact with colleagues helps to prevent an insular outlook and provides professional stimulation. For a similar reason the occupational health doctor or nurse should attend refresher courses from time to time. This is particularly important for those who normally work alone. Occupational health, like all other sciences, is developing rapidly in knowledge and understanding and it is essential for those

The Organisation of Health Services at Work 23

who practice it to keep up-to-date. This is also required of them and their employer by the courts if legal claims are made for failure to prevent occupational disease.

Specialist training courses in occupational hygiene are now arranged in some universities, but only relatively few as yet offer a comprehensive course covering all aspects of the subject. Applicants are usually required to have an engineering, chemical or medical qualification before being accepted for academic courses. Many more places offer short specialist courses on specific areas of the subject such as noise, radiation or dust. Training courses in ergonomics, or human factors as it is termed in the United States, are more prevalent and may be provided by university departments of applied physiology or psychology, although departments of ergonomics have recently been established in several universities.

First-aid

The true sense of first-aid is the carrying out of essential emergency treatment for illness or injury. The casualty is then passed on to a doctor, nurse or other skilled person for *second-aid.*

Unfortunately, the use of the term first-aid has also come to include the definitive treatment of minor injuries. No second-aid is envisaged. The treatment carried out is therefore unsupervised and without professional follow-up. A further difficulty is that the person attempting definitive treatment has to be certain that the condition is one which can safely and properly be treated in this way. Therefore, discrimination is required, and this may require more knowledge than the first-aider usually achieves. It is apparent, therefore, that different skills are required in carrying out first-aid and in providing definitive treatment for minor injuries. The essentials of true first-aid, as defined above, can be taught to intelligent people in about 8–10 hours, including practical training. On the other hand, discrimination in dealing with illnesses and injuries is a skill which can only be learned gradually by seeing large numbers of cases, by using the experience profitably and by studying the problems of discrimination.

In offices, factories and other similar industrial situations, there is a need—and usually a legal obligation—to provide for first-aid in emergencies. Proper selection of personnel combined with modern training methods can equip people fairly quickly with the necessary knowledge. Follow-up training is also required if the knowledge is to be readily available for use in emergency.

In small industrial groups, the definitive treatment of minor injuries is the part which presents greatest difficulty. Discrimination cannot be taught in a class except in a very general way for there is no adequate substitute for experience in this field. It is therefore wise to give an instruction to those who have to carry out this task that all but extremely trivial cuts *must* be referred elsewhere—to the patient's doctor, to a hospital or to the professional adviser retained by the management.

Most countries lay down legal standards for recognition of qualified first-aiders and for the amount and quality of equipment which must be available in relation to the number of people employed. If there are special hazards in any work place or area, it is essential that first-aiders be given special training in these hazards and in how to combat them. Standard first-aid courses deal with all common emergencies, but cannot be expected to take care of local and special requirements. In large organisations employing professional staff, whether doctors or nurses, the organisation of first-aid is best delegated to those people and not left, as is sometimes the case, entirely in the hands of old-time first-aiders. Experience shows that new and up-to-date ideas tend to percolate more readily through professional channels.

Further reading

Encyclopaedia of Occupational Health and Safety, International Labour Office, Geneva (1971).

Occupational Health Practice, R. S. F. Schilling (ed.), Butterworth, London (1973).

Annual Reports of HM Inspector of Factories, HMSO, London.

The Occupational Health Nurse, Occupational Safety and Health Series, No. 23, International Labour Office, Geneva (1970).

Occupational Health Nursing, Royal College of Nursing and National Council for Nurses of the UK Royal College of Nursing, Henrietta Place, London, W1 (1968).

Organisation of Industrial Health Services, Safety and Health at Work Booklet No. 21, HMSO, London.

The Nurse's Contribution to the Health of the Worker, Report of the Nursing Sub-committee 1966–1969, Permanent Commission and International Association on Occupational Health, c/o Royal College of Nursing (1969).

Essentials of Occupational Health Nursing, Doreen Pemberton, Arlington Books, London (1965).

New Essential First-Aid, A. Ward Gardner & Peter J. Roylance, Pan, London (1972).

The Fundamentals of First Aid, Robert A. Mustard, St. John Ambulance Association and Brigade, Ottawa (1955).

Occupational First Aid, St. John Ambulance Association and Brigade, Macmillan Journals Limited, London (1973).

CHAPTER 4

Routine Examinations and Screening Tests

All over the world doctors in occupational health perform routine medical examinations and for many this is their main activity. This does not, however, imply that it is necessarily the most useful function of an occupational health service. The procedure is expensive and on economic grounds alone should be subjected to cost effective comparisons with the other methods which may be used to assess health, and which are considered in this chapter. The reasons for doing examinations range widely from meeting legal requirements, ensuring that employees are fit for their work or have not suffered in health from it, to reasons that are so obscure as to suggest that this may be the only way that some people can think of using a doctor. Some doctors spend virtually all their working hours doing routine examinations on large numbers of healthy people; this blunts their ability to recognise the few abnormalities that may occasionally arise and the procedure becomes a meaningless chore. Perhaps the principles of quality control might suitably be considered under such circumstances.

The choice today is not simply between a medical examination by a doctor or nothing, for there are now many other choices to consider. These include health declarations, questionnaires and interviews, single and multi-phasic screening tests. Any of these can be used to supplement or partially replace the traditional medical examination by a doctor. There is also the elaborate medical check up, largely pioneered in the United States, which involves all of these plus admission to a clinic for a day or two for further investigations such as barium meals and sigmoidoscopy. With this wide range of choices there may be some confusion in the minds of doctors as well as of managers about their relative merits and values. The procedures have different objectives, involve different resources and most are more or less costly to perform, but their benefits can be a great deal more difficult to measure than is sometimes supposed. There are managers who believe that a pre-employment physical examination is the best way to reduce sickness absence, but as shown on p. 34, physical fitness is only one of many factors affecting absence attributed to sickness, and not a very important one. Many employees and their unions press for frequent routine medical examinations for all because they seem to believe that this is the best way to preserve good health but, despite its obvious attractions, this theory as yet lacks proof. In addition there is the problem of whether there are enough doctors available to do the work.

Reasons for medical examinations

Whether there is a legal obligation or not, there are two basic reasons why such examinations are done: (i) to ensure fitness for the job and (ii) to detect any damage to health that may have arisen from the job.

Ensuring fitness for the job covers a wide range of situations such as recruitment, acceptance by pension fund, transfer to other jobs, reassessment after injury or illness, and regular surveillance of people at special risk, for example the young, the old or the disabled. Detecting damage to health is more specific and is usually restricted to workers exposed to known or suspected physical, chemical or biological hazards.

Assessment of fitness for an occupation requires not only a knowledge of medicine but also an adequate understanding of the nature and demands that the job itself imposes on people. Thus, for example, the fact that a doctor may consider his patient fit to drive his private car after a heart attack, must not be taken to imply that he can drive a company vehicle to a tight schedule in heavy traffic. It is, therefore, necessary for a doctor working in industry to familiarise himself with the work place and its processes, and also to understand any local productivity arrangements that may involve flexibility both of work and possibly also of working hours. That bane of every supervisor's life, the 'light duty' certificate, is another example of the need for relevant medical advice from a doctor or occupational health trained nurse, who knows enough to make constructive and realistic proposals on what sort of work the individual can manage.

Fit or unfit for work?

A medical examination, particularly at recruitment, has but a limited value for the prediction of subsequent health, and even less value in the production of subsequent absence from work attributed to ill-health. A reliable crystal ball is, unfortunately, not part of a doctor's stock-in-trade. A recruitment or other medical examination to assess fitness for work should ideally match the individual's physical, mental and psychological abilities on the one hand against the corresponding demands of the job on the other. Even with unlimited funds and a fully equipped medical and ergonomic laboratory, such matching would still be imprecise. Normal occupational health services manage without such facilities. The usual type of examination does no more than find out that the subject is, or is not, *unfit* for the work at the time of the examination and, depending upon the nature and severity of any condition discovered, whether the situation is *likely* to improve or deteriorate in the future. It might even be advisable for medical certificates to be worded in this way rather than in the conventional 'fit for work' manner.

Examinations and other techniques

Medical examinations carried out in the traditional way by a doctor have three stages: (i) oral questions and answers ('the medical history'), (ii) physical examination ('laying on of hands') and (iii) measurements and other investigations ('tests'). The term *'medical examination'*, although often used to describe all types of health checks, is more properly restricted to one performed by a doctor that includes a comprehensive physical examination. The entire procedure of the traditional consultation is often done by the doctor himself, often with the aid of a standard proforma. It is now common practice, however, for some standard questions to be asked by a nurse assisting him, who also undertakes the measurements of height, weight, vision and the testing of urine. The results may sometimes reveal a need for additional tests such as an X-ray or a blood count. In other circumstances these may be done routinely as part of the examination of people such as food handlers, radiation workers and nurses. Health declarations can also include an *occupational history*, because this is of great importance in assessing suitability for work. For example, previous exposure to dust hazards or radiation may preclude further potential exposure.

The other techniques described in this section concern modified versions of the interrogation or test phases of the full consultation. A *health declaration* or self-completed health questionnaire has been used by life insurance companies for many years. It has now been adopted by many large organisations—notably those with high labour turnover—as a standard procedure for recruitment. The two great advantages are speed and economy. The completed questionnaires can be sorted by lay clerical staff if carefully designed. About 20 per cent may need to be passed on to doctors or nurses for advice on the need for medical examination. About half of these are usually recommended for a full examination, but the advantages in economic terms are obvious. Although such questionnaires can be answered untruthfully, a section giving consent for information to be revealed to the company's doctor by the applicant's family doctor, can act as a deterrent when coupled with a warning on the possible consequences of concealment. The experience of organisations who have tried this system suggests that deliberate concealment of important conditions is uncommon provided the questions are clearly and carefully worded.

Intermediate in position between a health declaration and a full medical examination are a *health interview and tests* performed by a nurse who has been trained in the technique. A standard series of questions are asked with supplementary questions as necessary. Measurements are made of blood pressure, pulse rate, height, weight, vision and a specimen of urine is examined. Here, too, there remains a need to refer a small proportion, about 5 per cent, for a medical examination. Much of the information the doctor needs will already have been obtained and thus his time will be saved. This arrangement is perhaps the most satisfactory in cases where the work involves health hazards to the employee, or in the goods or transport industry, where there is a potential risk to the general public. The procedure usually permits the nurse to pass an applicant as fit, but requires a reference to a doctor for those who may be unfit. There is also an undoubted economic advantage compared with full medical examinations of everyone.

Screening tests are now being more widely used as an effective and economic means of searching for one or more specific abnormalities in groups of apparently healthy people. Mass miniature radiography (MMR) is perhaps the best known procedure which has been used all over the world to detect tuberculosis and other chest diseases. Other tests in widespread use include cervical smears to detect pre-cancerous conditions of the cervix, and urine tests on newborn babies to find the rare but treatable biochemical disorder of phenylketonuria. In occupational health practice the screening of electricians, pilots and deck officers for colour vision, and of people in noisy work places for hearing has now become quite common. Tests for the effects of toxic substances such as solvents, lead or insecticides can also be used to ensure that environmental hazards are adequately controlled at the workplace.

The development of automated methods of chemical analysis on small quantities of blood have allowed *multi-phasic screening tests* to become more widely used since unit costs have fallen. The procedure takes but a few minutes of the subject's time. Many chemical and other constituents of the blood can be tested automatically within a few hours. The results of this battery of tests, sometimes called a biochemical profile, are then passed to a doctor for interpretation and decision on any further action necessary, although some of these lists of analyses can be unnecessarily lengthy. Thus the results will often be of little direct value except, perhaps, as a baseline for discovering subsequent changes. Most laboratories offer a package which may include some unwanted items. Although unit costs of tests may seem to be low, the total cost may be appreciable.

The last procedure to be described is a combination of screening tests and a medical examination for *pre-symptomatic diagnosis*. The underlying objective is convincingly simple, since it is based upon the principle that detection of disease at the earliest possible stage should lead to more efficient treatment and thus to the preservation of health and the saving of life. This form of screening was first developed in the United States and is now most elaborately and comprehensively performed in clinics designed specifically for the purpose. The cost is substantial. Clientele comes mainly from the ranks of senior executives. The analogy with preventive maintenance of vehicles is frequently used in the advertising of such a service, as are statements such as that the commonest cause of death in executives is coronary thrombosis, and its risks can be reduced if certain chemical or biological abnormalities can be detected and corrected in time. We shall return shortly to the validity of such claims. The entire procedure, however, combines all the different techniques described of self-completed questionnaires, medical examination and multi-phasic screening tests.

Medical ethics and confidentiality

As the procedures considered in this chapter mainly fall outside the traditional practice of treatment-orientated medicine, the usual doctor-patient relationships do not apply. This difference is most obvious in recruitment examinations and in those medical examinations required by law. The role of the doctor in a pre-employment examination is similar to that in examining an applicant for life assurance. The results must be made known to the employer and the outcome is presumably of some importance to the applicant. The doctor's role is only that of medical examiner. He is expressly excluded from offering treatment, although he should inform the patient's family doctor of any medical condition that he may find.

In the case of voluntary examinations the doctor operates under the usual constraints of confidentiality imposed by ethics and by law. Even here, however, the normal doctor-patient relationship is not fully developed. Similar situations arise in screening tests, and can be at their most difficult when pre-symptomatic diagnostic check-ups are offered. When these are requested by the employer the results may also be made available to him, even if only in general terms. More frequently, however, the employer pays the bill without requiring to have specific reports. This arrangement is certainly preferred by both doctor and patient. The doctor can, when necessary, obtain the patient's consent to allow him to make a non-clinical report to the employer.

The merits or de-merits of screening tests

Regardless of who receives a report, these examinations are in effect being offered as a service with the implication that subjects who participate are likely to derive benefit. This is where a number of the most difficult problems arise. Some conditions discovered in this way may well be curable. Early tuberculosis of the lungs detected by MMR is a good example. Many other abnormalities that may be found are less amenable to cure. Even in the days before a specific cure for tuberculosis had been found, its early detection and treatment certainly reduced the risks of infecting others. This seldom applies to other conditions likely to be found with pre-symptomatic screening tests as the rise and fall in the extent to which MMR facilities have been provided in several countries makes clear. As long as tuberculosis remained a nationwide scourge as it used to be in Western Europe, the extensive use of MMR was fully justified, but in those places where the disease has become rare its cost is no longer economically justifiable. Certain places remain, however, where it is still provided, such as air- and seaports, and towns where immigrants are found. The provision of MMR facilities in places other than these results in few new cases being found. The discovery of other diseases such as cancer of the lung—which rarely responds well to treatment—does not justify the continuation of the service. If and when an effective cure for lung cancer is found the policy may well be revised.

If, therefore, the subject who volunteers to take part in a pre-symptomatic diagnostic check up is found to have a certain condition, it would be reasonable for him to expect that something effective could be done about it. This provides the problem which has caused so much controversy. Contrary to general belief, the life-saving or disablement-preventing value of many of the screening tests now used has yet to be scientifically proved. Large-scale trials are now in progress in many countries all over the world to study this very point. Take, for example, the discovery of high blood pressure in a subject at such an examination. There is ample evidence already to show that if the pressure is very high indeed, treatment with appropriate drugs can improve his expectation of life, which would otherwise be poor. As yet there is little evidence whatsoever to show that the same would apply if the pressure were only moderately above the normal. Indeed, there is still disagreement on the levels at which normality ends and abnormality begins. Treatment, which it is not the responsibilty of the diagnostic clinic to prescribe or supervise on a long-term basis, can itself have uncomfortable side effects. A large scale trial is now taking place to find the answer and these clinics are well-suited to take part. This is important since between 5–10 per cent of apparently healthy adults of middle age have a raised blood pressure. A similar dilemma may be found in relation to factors known to be *associated with* coronary thrombosis. Here, too, trials are in progress to evaluate the benefits of drug or dietary régimes. All agree that research is urgently needed, but in the meantime it would be immoral to allow volunteers to believe that they will, with certainty, benefit. It is also questionable for some to imply that coronary heart disease is a condition for which executives are peculiarly at risk. Existing evidence from the United States and the United Kingdom shows that men in the lower ranks of the factory hierarchy are even more at risk.

It is very sad that the only time some patients may ever feel that a doctor can spare time for them and really listen to their problems is in the course of a planned routine or periodic medical examination. The relief which many patients feel after hearing that no abnormality has been found and the sense of having 'got things off their chest' when they were listened to are both very real. Whether the amount of time and effort is justified, whatever that may mean, on this basis remains questionable, so the merits of screening procedures have been argued at length in the medical and lay press for years. The critics draw attention to the risk of encouraging hypochondriasis and generating unnecessary anxiety when minor abnormalities are discovered. They have doubted whether many conditions found can be effectively cured or even controlled, and finally warn that if nothing amiss has been detected that reassurance given can encourage people to ignore subsequent symptoms. Those who support screening feel that they are pioneering a growing point of medicine. They use the analogy of preventive maintenance and make the points that advice on personal preventive medicine is a useful service, that some conditions can indeed be cured or controlled, and that procedures which include an interview with a doctor allow the patients to seek advice on other, perhaps minor, conditions which worry them but for which they have not yet sought the advice of their family doctor.

In our opinion such screening tests do have some real value when coupled with a medical examination, but we also believe that it is essential that their limitations are clearly appreciated. The family doctor must be advised of any significant abnormality detected, for it is he who will bear responsibility for the long-term treatment of the condition. Finally, for the employer who may have to pay for such a service, we would point out that the detection of one case of curable serious disease in a senior member of staff may, by the use of discounted cash flow calculations, be shown to save more over ten years than the cost of such examinations for perhaps a hundred senior staff. The need, in terms of death and illness prevention is not however restricted to such highly-paid employees. The argument to restrict the service should be on financial not medical grounds.

Biological monitoring for toxic hazards

Screening tests may be required to identify employees who have absorbed significant amounts of toxic chemicals or whose health may have been affected. Another use of screening tests is as a supplementary procedure to show that the working environment is safe, since clearly it would be wrong and dangerous to wait until poisoning occurred before making use of environmental hygiene measurements. An abnormal result from a screening test on an employee means that the safe working environment control procedure has failed. Routine screening tests are now required by law in many countries to ensure that certain hazards such as lead are well-controlled. These sort of tests are a proper and necessary function of any occupational health service.

With substances such as carcinogens that may take years to cause disease, the problem is somewhat different. For example, the risk of bladder cancer arises some twenty to thirty years after exposure to the cancer producing substance. Regular testing of urine specimens from people known to have been exposed to the hazard can reveal tumours in an early and treatable stage. Even when the condition is irreversible, there may still be benefit to other workers if a hazard can be shown to have existed since this evidence may lead to more effective control. In all forms of biological monitoring for occupational hazards it is necessary for the doctor to know the nature and degree of exposure. For instance, in one case where a man who had occasionally worked with asbestos was examined, it transpired that as an amateur steam engine enthusiast he had spent many hours of his spare time stripping blue asbestos, the most dangerous kind of asbestos, from around the boilers of his beloved engines. Fortunately, no abnormality was found at his examination. The medico-legal position would, however, be interesting were he to develop asbestosis, a lung tumour, or a mesothelioma (p. 81).

In summary, there are many methods which may be used to check upon the health of workpeople. The most efficient method should in each instance be selected to meet the local need. It is difficult to quantify benefits in the profit and loss account.

Further reading

Principles and Practice of Screening for Disease, J. M. G. Wilson & G. Jungner, World Health Organisation, Geneva (1968).

Screening for Health: theory and practice, H. P. Ferrer, Butterworths (1968), London.

Effectiveness and Efficiency: random reflections on health services, A. L. Cochrane, The Nuffield Provincial Hospitals Trust (1972).

Mass Health Examinations, Public Health Papers No. 45, World Health Organisation, Geneva (1971).

CHAPTER 5

Absence : Sickness or Malingering?

It is said that people stay away from work for one of two reasons—either they are prevented from attending, or they choose not to attend. However, this oversimplifies the problem since it presents the black and white but ignores the many shades of grey that are far and away the most common. Absenteeism, and especially absences from work attributed to sickness, has attracted a lot of attention over the years. The subject seems to be one that encourages many to air their prejudices and to oversimplify to the point that sometimes verges on the absurd. Statements such as 'It's all because doctors are too free with sick notes,' or 'the main cause of sickness absence is malingering', may often be heard where two or three managers are gathered together in almost every industrialised country throughout the world. They tend to be followed by proposals for remedies that have a similar ring to them: 'cut sick pay rates and absences will drop' or 'a million unemployed will cure the problem'.

This chapter will begin by considering some facts about the nature and causes of 'sickness absence' or, as it is more accurately described, 'absence from work attributed to incapacity', before making suggestions for those who would like to design their own control programmes.

Most organisations keep records of absence of attendance, if only to allow wages to be calculated correctly. The reliability of such records however is usually inversely proportional to the status of the staff concerned. Rarely is the absence of top management recorded in full—and there may well be occasions when work is indeed done on the golf course. The criteria of what shall or shall not be included as 'absence' vary widely from one firm to the next. Some include annual holidays or company-sponsored training, most include absence attributed to incapacity, so-called 'sickness absence'—with or without a supporting medical certificate—and casual absence for non-medical reasons. Absence for personal or family reasons which has been given prior sanction may sometimes be included as well. The importance of these other causes of absence to this discussion is that the relative proportion of 'sickness' compared with the total amount will depend in part on what is, or is not, included in the total. Thus in a country with, for example, two weeks military reserve training for all men under the age of forty, an organisation that includes this as well as annual holidays as 'absence' will be less concerned with an average of two weeks of 'sickness absence' than one that restricts its definition of absence to include sickness and casual absence only.

Before attempting to compare 'absence' rates between different companies it is as well to enquire what types of absence are included in each company's figures. Those having sick-pay schemes covering the first or single days tend to find that 80–90 per cent of all absence is due to sickness, while those with no sick-pay schemes at all record only about 60 per cent. Such an observation does *not* justify the conclusion that the difference is due to malingering. With no sick pay for absences of a day or two there is little incentive for the employee to obtain and submit medical evidence of sickness, and where the illness lasts for a week or two there may be little advantage to him in showing a medical certificate to the employer. Experience has shown that firms having strict absence-control procedures, coupled with disciplinary sanctions for unauthorised absence, tend to have more absence attributed to sickness than those who do not, their work people taking 'sickness absences' of a week instead of unauthorised absences of a day or two.

The meaning of medical certificates

The production of a medical certificate is not necessarily proof that the individual is completely incapable of work. Serious illnesses or injuries that are incompatible with work of any sort form but a small proportion of all episodes or spells of sickness absence in any organisation. Whether one approves or not, it must be appreciated that in virtually every country today it is the worker himself who is usually the one who decides if and when he is going to stay off sick and, at the end of an illness, just when he will return. There is a vast amount of ill-health in every community even though the bulk of it is minor or trivial. A recent study of 2000 adults in South London showed that while only 8 per cent had lost some time from work during the two weeks before the interview, 20 per cent had consulted a doctor and 95 per cent had suffered from some symptom or other. Most people, therefore, continue to go to work despite trivial or minor ailments. However, if a man or woman with a backache or a headache feels unable to go to work he or she will not find it unduly difficult to persuade a doctor to sign a medical certificate.

Just as frequent is the difficulty which the doctor has in deciding when someone is fit to return to work after an illness or an operation. Social security systems in almost every country require a sudden transition from incapacity to fitness from one day to the next. Many employers take the same view, so temporary modification of work appears to be well-nigh impossible in work places. Convalescence is always gradual and, from a medical viewpoint, there should be a gradual increase in work from soon after the acute phase of the incapacitating illness is over until full recovery has taken place. The doctor who signs certificates often knows little or nothing of his patient's job and less about how it may be modified. Unless there is an industrial doctor to whom he can turn, he is forced to take the advice of his patient and on some occasions, at least, this advice may scarcely be called objective.

This brings up again the role of doctors and their responsibilities when issuing certificates. Even though they may have been given the responsibility for advising the social security or sickness insurance officials about incapacity and also, less directly, the employer about their patients' fitness for work, they consider that their primary responsibility must always be to their patients. One of the problems in medicine, as in many other subjects, is the difficulty in proving a negative. Thus although a 'shocking backache' or a 'splitting headache' is probably of no great medical importance, it might possibly be an early symptom of serious disease. A large part of the medical curriculum in every country consists in teaching students how to recognise the early signs and symptoms of serious illnesses. An important part of this consists of taking a 'case history' from the patient since many conditions produce symptoms before any objective physical evidence can be detected. This training therefore means that doctors must necessarily start by accepting their patients' statements at face value, and is essential if a satisfactory doctor–patient relationship is to develop.

Most doctors in industry would agree that pure malingering is uncommon, though it is occasionally seen, but that hypochondriasis or the elaboration of minor ailments is a much more substantial problem. The man with a chesty cold who oversleeps on a rainy morning has indeed got a cough and a cold; who can prove that he is really not fit enough to go to work? Doctors who fail to recognise serious disease risk their reputations and perhaps court action. A prescription and a chest X-ray, together with a few days off work, might prevent something far more serious. The old fable of the boy who cried 'wolf' ended in tragedy. Even when the doctor may suspect that there is little or nothing wrong with his patient, he must know him very well before deciding to ignore his statements completely.

Once it is accepted that workpeople themselves exercise a substantial element of choice in the decision to absent themselves from work—whether attributed to sickness or to other reasons—the possibilities of successful management action become apparent. Some of the most useful clues can come from studying those who seldom go off sick. A common misconception is that sickness strikes at random; this is certainly not true and still less the idea that sickness absence is 'the luck of the draw'. So called 'sickness absence' is not of course a good measure of sickness, it simply represents the absences *attributed to* sickness. Thus it is only partially dependent upon the amount of disease in the working population. This, too, is an essential point to appreciate if a realistic absence control programme is to be drawn up.

The responsibilities for controlling absences

A reduction in absence attributed to sickness can only be achieved by the combined action of two separate but interrelated activities. The first is a reduction of ill-health in the working population—whether it be due to local environmental factors at work discussed elsewhere in this book or to diseases affecting the whole community such as influenza, bronchitis, heart disease and so on. The control of disease as such is primarily a medical responsibility, but one that is only likely to be effective if it has the participation and support of management, staff and unions in the work place, or of government and the general public in the whole community. The other way of reducing absence attributed to sickness is to reduce the absence rather than the sickness. This is a managerial responsibility but this, too, is likely to be more effective if it has the assistance of specialists in various fields including those in occupational medicine. It may be relatively easy by administrative action to alter the form that absence takes—for example, a punitive policy for lateness coupled with generous sick pay for single days off will reduce lateness at the expense of days off. Indeed, it is most surprising how many organisations do just this. The system of 'waiting days' with benefit paid from the first day if the absence lasted longer than three days certainly encouraged both patient, and sometimes the doctor, to lengthen short spells. Organisations whose records show virtually no casual absence should look at single days of 'sickness', and so on. *To be effective, any absence-control programme must cover all forms of absence.*

Most organisations do have some procedures to deal with absence, but they often concentrate upon lateness at one end of the scale and very long sickness absence at the other, ignoring what is probably the biggest problem today, high-frequency absences of relatively short duration. The control of absence is ultimately the responsibility of line management, but before designing a suitable programme the manager must identify the main problems within his own firm. This requires a certain basic amount of relevant information about how much absence

Absence: Sickness or Malingering?

there is, what form it takes, and in which parts of the organisation it is most frequent.

The measurement of absence

The commonest measurement in use in industry is undoubtedly the 'lost-time percentage'. This shows hours or days of working time lost, expressed as a percentage of total potential working hours or days that could have been worked had everyone attended for their normal duty. This excludes the extra hours of overtime which, were they to be added to the denominator of the sum, would depress the rate. Two other measures are also frequently found: the 'absentee percentage' which expresses the number away on a day as a percentage of the total number due in to work that day, and the 'sick rate' which expresses days lost per annum as an average number per person employed.

Absence is, alas, a repetitive event which can vary both in frequency and in length of spell. This means that in contrast, for example, to units of production or to births and deaths where single measures suffice, absence *must* be measured in terms both of frequency and of time for the information to have meaning. The inadequacy of using a lost-time rate alone is easy to illustrate. If, in a section of ten men, one was absent with a serious illness for a very long time, the lost time rate would be 10 per cent. If, on the other hand, there had been numerous unpredictable brief absences involving all the men, the lost time could also be 10 per cent. The problems faced by the supervisor in the two examples are quite different. This would not be shown unless a measurement of the number of separate spells of absence—a *frequency rate*—was also made. This is probably best presented as an average number of spells per person measured over a period, usually a year.

Rates often encountered in industry may be in the order of 5–10 per cent of potential working hours (lost time) with a frequency of between one to two spells per person. Such averages conceal wide differences between men and women and also between blue- and white-collar workers. While generalisations are risky, a factor of between two and three often separates these main groups. Thus male white-collar workers may have about 2 per cent lost time and male blue-collar workers about 6 per cent. Women, with the exception of career women, tend to have about two or three times the rate of men. Similar differences have been shown between regions. Men in Wales for example have about three times the amount of absence attributed to sickness as men in East Anglia and in the South East region of Great Britain. These sort of differences have also been described from many other countries and should always be considered before comparing rates between work units in different locations. Even within one location, departmental rates can vary widely. Measurement of these will allow the manager to concentrate his attention in the most appropriate place.

The likelihood of presence and of absence

Absence, like all forms of human behaviour, is neither evenly nor randomly distributed in any group of people; a few will have a great deal of absence and some will have little or none. There is substantial evidence from many countries and types of organisation that about 5 per cent of any workforce accounts for about 30 per cent of all spells or episodes and a further 5 per cent for about 30 per cent of all days. These two small groups usually consist of different people: the 'repeaters' or high-frequency absentees are mostly younger people, but the long-duration absentees tend to be rather older and many of them have more serious conditions. Any plan for absence control must ensure that these individuals are readily identifiable; some may prove to be physically or mentally unfit but others may have a remediable condition. The criteria used to identify both repeaters and long duration people must clearly be related to the locally prevailing rates, amongst men and women, both blue and white collar. The high frequency absentee is usually the most troublesome from the manager's point of view. He also tends to have other characteristics: he may frequently be late, takes more casual absence, has more attendances at the works medical centre and suffers more minor injuries. Work performance when he is present may be either good or bad, so dismissal is not necessarily the wisest course of action. Many have social or personality problems underlying their behaviour and are often frustrated or disgruntled at work. Few show signs of significant disease although some may have a drinking problem. There are a number who, when they have been transferred to other more responsible jobs, change out of all recognition and become models of good attendance. Procedures to deal with high-frequency absentees must be flexible if injustice is to be avoided.

The underlying causes of absences

A careful investigation of any episode of absence, irrespective of its apparent cause or diagnosis, may show that its underlying cause may have been due to the job, to a purely medical condition, to a social problem at home, or to a behavioural difficulty in the individual's personality—or even by any combination of these. For example, a day or two off for a headache, with or without a medical certificate, might really be due to any of these. The possible occupational causes could range from a bang on the head, mild carbon monoxide poisoning or a stressful relationship with colleagues. Purely medical causes, which are less frequent than is commonly supposed, might range from sinusitis to

Absence: Sickness or Malingering?

a brain tumour. Social causes might include marital discord or the sufferer's child in trouble with the police, and behavioural causes an inadequate personality or pure malingering.

Although it is often true that short spells of absence attributed to sickness tend to have more of a 'social' than medical content, those who are reluctant to go sick at all may only take a day or two off with influenza even though most people are more inclined to take a week or two. At the other end of the spectrum, lengthy absences tend to have a substantial medical content, but even here the timing of return may be influenced by motivation and other factors. It is necessary to look at the absence pattern of the individual rather than a single episode. It is in this way that 'repeaters' or the long-duration absentees should be identified. A record showing twenty-four days off in *one* spell in a year is of quite different significance to one showing twenty-four days in *nine* spells. In all probability the former was due to a medical condition and the latter, despite the label 'sickness', to predominantly non-medical reasons.

The never sick

In contrast to those who seem to take time off work for the slightest excuse, are others who seem to be immune to the calls of absenteeism. Every organisation will be found to have its quota of people who never take a day off sick for years on end. Although many of these 'never sick' are found amongst the ranks of management, about 5 per cent of blue collar workers will also behave like this. One study of such men showed that they were, in their way, just as unusual as the frequently sick when compared with the main bulk of their colleagues. They were rarely or never late for work and seldom had accidents or, to be more precise, they never reported having any. They avoided the first-aid or medical centre and they rarely saw their family doctor.

Their personal attributes and medical background provide some clues on how to select such people. They are usually over the age of thirty and claim to have had a wonderfully happy childhood (despite factual evidence in some of family tragedy or bereavement) but marital discord is not uncommon amongst them. They will have had an excellent attendance record in their previous job, and once employed will be content and not hanker after promotion. The medical standards of recruitment may appear unusual, since bad dental hygiene with teeth that have clearly never received dental attention, and perhaps a chronic physical disability, need not deter those who recruit. They will, however, have no history of nervous trouble, backache or bowel disorder. They may also show a marked preference to using a bicycle to get to work even though they own cars. These people, once employed, will not necessarily be the best of workers, but their attendance record will be impeccable.

The place of recruitment medical examinations

On the principle that prevention is better than cure, many organisations have arranged for pre-employment medical examinations based upon the mistaken belief that sickness absence can be predicted in this way. The value of these examinations is discussed elsewhere (p. 26), but they serve no useful purpose in this particular respect. One study of three groups of men, the first with high-frequency short absences, the second with long-duration absences and the third with none at all, showed quite clearly that even with hindsight the comprehensive physical examination they had all passed on recruitment revealed nothing of predictive value. Almost all the men in the high-frequency group had remained physically fit in objective terms, but about a quarter of the 'never sick' were found to have significant physical disease. One, for example, had been recruited as a 'disabled person' since he had been partially paralysed by a war-time head injury and he could only walk with the aid of two sticks. He had been employed as a clerk in the stores for fifteen years without a single day off sick.

We can only repeat that so called 'sickness absence' is first and foremost absence and has only a secondary link with illness. Within reasonable limits the objective physical condition of a man matters little, what does matter is his attitude towards himself and his health, and his motivation to attend regularly for work.

Predicting absence

How then may absence be predicted? The only method proved to be reasonably reliable is the individual's previous record of absence from work. A six-month period is usually sufficient, but is not completely reliable. As we have already mentioned, a 'repeater' allowed a change to a job which improves his motivation, can reform virtually overnight. Nevertheless, we believe firms could make more effective use of trial periods after first engagement, during which frequent spells of absence, however labelled, should not be allowed to pass unheeded.

Controlling absence

Specific measures to control absence can be considered in two groups: first prevent the absence commencing and second, reduce its duration.

Prompt and effective first-aid treatment for minor injuries and, where practicable, perhaps prompt treatment of minor illnesses or reassurance of anxieties about health, can reduce the need to visit hospitals or family doctors. Preventive health procedures such as immunisation can protect against a few specific infectious diseases—but the case for influenza immunisation on a large scale in industry is still open to question both on medical effectiveness and on economic grounds. Most firms that have tried it find that the numbers who volunteer tend to drop off after the first year or two. Unfortunately, no effective vaccine is yet available for coughs or colds. An appropriate personnel policy and sick-pay scheme is clearly essential and the personal interest of line management at all levels up to and including chief executive is essential. Strange as it may seem, we are convinced that this needs to be said since too many line managers or supervisors opt out of this responsibility and are happy to see it done—less effectively—by personnel staff. The return to work after sickness can also be speeded-up if the absentee is encouraged to return as soon as he feels able to do so, and not to wait for the return of 100 per cent fitness, which is at best a largely mythical concept. Where a doctor is available at work he can often, by developing a good liaison with local doctors, speed the return to work by explaining job requirements and, on occasion, by arranging temporary modification of duties or hours to enable return to take place as early as possible.

The traditional day for returning to work after a long absence is Monday. Consideration of the problem will show that the best day to return is Thursday because this allows a more gradual resumption of work, starting with two days at work followed by two days off and then five days at work. A similar principle should govern resumption of shift work after a long absence.

The second-century physician Galen, is credited with the aphorism 'work is Nature's best physician'. This is not less true today. Motivation, however, provides the key and a person who feels that his or her work is useful and needed is unlikely to take more time off than is really necessary. Incentive schemes involving good attendance bonuses are sometimes said to provide a solution. In fact, great caution should be exercised before embarking upon them for they tend to lose their beneficial effect fairly rapidly and the demand that they be continued indefinitely may prove difficult to refuse.

Absence attributed to incapacity is first and foremost a problem for managers, and the role of medical or nursing advisers must be viewed as an aid to good management not as its substitute.

Further reading

Absenteeism, An Analysis of the Problem, Confederation of British Industry, London (1970).

Absenteeism — Causes and Control, P. J. Taylor, The Industrial Society, London (1973).

Absenteeism, R. M. Jones, HMSO, London (1971).

Absence from Work attributed to Sickness, A. Ward Gardner (ed.), Society of Occupational Medicine, London (1968).

Off Sick, Office of Health Economics, London (1971).

Sick Pay, E. G. Rutter & K. Ottaway, The Industrial Society, London (1967).

On the Quality of Working Life, N. A. B. Wilson, HMSO, London (1973).

CHAPTER 6

Disease, Impairment and Disability

Most people with disabling conditions are very reluctant to describe themselves as 'disabled'. There are a number of reasons for this attitude, but one of the most important is the widely-held belief that a label such as this can cause serious problems in obtaining or keeping a job. Unfortunately this fear is often only too well-founded; the whole subject is often clouded both with emotion and by an inadequate understanding of what is really involved. Even the word 'disabled' itself is used with different meanings by different people, and on some occasions by the same person in different contexts. To many the word 'disabled' seems to conjure up a picture of a mutilated war veteran who may be substantially, if not completely, unemployable. This is, of course, only one very small part of a problem which ranges widely both in cause and severity. Just as some of us may use the words 'accident' and 'injury' as if they were synonymous, which they certainly are not, so in the case of disability we must distinguish between the disability and the impairment that gives rise to it. It is also essential to appreciate the person's abilities so that what he can do receives as much attention as what he can not.

Medical conditions, impairments and disablement

It may be convenient to consider a sequence starting with an injury or a disease which we can call a medical *condition*. The problem may be illustrated by two examples: the loss of an eye, and angina caused by coronary artery disease, Such medical conditions, depending upon their severity, produce the second link in the chain, an *impairment* of bodily function. For the man with one eye the impairment is loss of binocular vision resulting in monocular vision. For the person with angina the impairment would be a reduction in the amount of physical exertion possible, described in medical terms as a diminished exercise tolerance. For an impairment to become a *disability* there must be a loss or reduction in the ability to undertake certain activities, such as work. Thus the one-eyed man could be disabled if he happened to be a crane driver where stereoscopic vision with depth perception is essential, but he would not be disabled for most ordinary jobs. The man with angina could well be disabled for a manual job, but could probably work at a sedentary one. It is clear that the word disabled should not be used alone, but should always be qualified as 'disabled for —'. By the same token there can never be an absolute condition of disablement. The most severely disabled are unable to care at all for themselves and need constant medical and other attention. A rather larger number are only able to cope with daily living with some assistance in the home, but far more are disabled only for certain activities which may affect their ability to work in some occupations. It is with this last group that we are concerned in the context of this book. Many of them are employed, but there are many others who have failed to get a job because of prejudice or fear on the part of those who could employ them.

Forecasts

With any disablement, the future outlook of the condition causing it—known in medical jargon as the prognosis—must always be considered. A static condition, such as that resulting from an injury or a congenital defect, is likely to affect long-term employability much less than a condition that is progressive. Muscular weakness of an arm or leg due to old poliomyelitis is easier to live and work with than a similar degree of weakness caused by a progressive disease such as multiple sclerosis. It is also not surprising that people find it easier to adapt themselves and compensate for a static condition, although even here there are wide differences between people in their ability to do this. As a general rule the young are better able to adapt than older people, and the capacity of children to adapt to what seem to be appalling impairments is well known. The victims of the thalidomide tragedy have shown this to the world.

One problem that should also be recognised is the difficulty doctors find in trying to make an accurate prognosis. Actuaries and insurance companies manage well with the use of statistical probability tables, but when it comes to individuals the chances of one person's chronic condition deteriorating can rarely be expressed in precise terms. Thus, although a man of fifty-five with chronic bronchitis and emphysema who has already had to take a month off work in each of the past two winters is very likely to have even more time off next winter, and may well become quite incapable of work within a few years, this is only very probable and any one man might not deteriorate as fast. The art of medical prognosis is not the exact science that many people seem to expect.

Disease, Impairment and Disability

Misinformed anxieties of managers can and do cause unnecessary restrictions to be placed on the employment prospects of people with conditions such as arthritis or diabetes. The word 'arthritis' is taken by many to mean the progressive general inflammatory disease of the joints called 'rheumatoid arthritis'. In practice, however, arthritis is more commonly used to describe the much less serious condition more properly called 'osteo-arthritis' which usually affects only one or two joints and is due to the combined effects of age and injury. Many well-controlled diabetics, too, have been refused employment in jobs such as a bank clerk, where their disease would not affect their ability to give many years of useful and productive work. This point may seem self-evident to most people, but there are a disturbing number of diabetics who have been refused jobs for reasons that appear to be based on prejudice. Doctors working in industry have an important responsibility to recognise such misinformed fears and to correct them as far as possible.

In certain jobs, particularly in transport, continued employment of people with a poor health risk is considered hazardous for the safety of others. Licensing authorities take an understandably cautious attitude towards the standards of health required of drivers of buses or heavy goods vehicles, and even more in respect of aircraft pilots. The fact that a pilot with a high blood pressure may feel quite fit for work today is rightly considered irrelevant in deciding on his fitness to continue flying passenger aircraft in the year to come. More common still is the problem of deciding on fitness to continue driving smaller vehicles at work after an illness, such as a heart attack. Here one must consider the safety not only of the driver, but also of the public. The risks will depend partially upon his health and also upon the amount of time actually spent at the wheel, the size of the vehicle and the type of route undertaken. Undue caution may result in many being deprived of their livelihood, but a reckless disregard of the risks may occasionally have serious legal consequences. The implications in each case will need careful consideration.

There remains that happier form of impairment and disability due to a condition that is likely to improve or recover completely. The discovery of pulmonary tuberculosis used to mean months or years of incapacity, but in most cases discovered today recovery and a return to normal work can be expected in a matter of weeks, even though continued medication may be necessary for months. As in all biological phenomena there is a 'grey area' between what can reasonably be considered a temporary impairment and one that continues for long enough to cause serious employment difficulties. The distinction between temporary and longer lasting impairment may have to be an arbitrary one.

Degrees of impairment

When a medical condition improves to full recovery, the impairment gradually reduces, and the actual day on which full recovery takes place is impossible to measure. However, in the administration of social security benefits and in decisions about fitness for work, the requirement is for a sudden transition from incapacity on one day to fitness on the next. As discussed on p. 35 in relation to absence attributed to sickness, this is a situation that allows the 'sick' person a wide element of choice in deciding for himself when he is going to return to work. It is unfortunate that neither governments nor employers seem to be prepared to recognise a state of partial recovery involving fitness for limited work and limited benefits. A similar problem arises with a decision on permanent or long-term disablement for social security benefits. Some countries require that the medical condition and its associated impairment should be described as 'permanent'; most require that the impairment is likely to last for six months or one year; a few set a minimum period of three months. Clearly, a manual labourer who has had an operation for the repair of a hernia and is unable to undertake heavy work for three months, should not be labelled as 'disabled', although a clerk may be able to return to his normal duties after a month. We believe that it would be reasonable to consider one year as the limit of temporary impairment. There will, however, always be some people whose condition improves sufficiently after that time to provide exception, and with developments in medical science these exceptions may become more common in the future.

Rehabilitation and resettlement

Rehabilitation is the process of restoring as much function as possible, and its purpose is to reduce the degree of impairment. *Resettlement* describes the process of finding suitable work for someone whose impairment has disabled him from returning to his previous occupation. This often involves retraining for new skills. For example, a miner may be disabled from his job due to an injury to his lumbar spine that impairs him permanently because he is unable to bend his back or lift heavy weights. In the weeks after the accident he will require rehabilitation to ensure that he is able to move his back as much as is possible and that he learns how to make the most use of his hips and the upper parts of his spine. He will then need to be resettled into another occupation and this, for example, could involve *retraining* as a coil winder so that within a few months he is able once more to earn his living at his new trade.

Rehabilitation is one subject about which all medical authorities agree as being of the greatest importance, but alas they seldom practise what

Disease, Impairment and Disability

they preach. This fault is not restricted to doctors since the need for good rehabilitation services is acknowledged by politicians, economists, health service planners, employers, leaders of trades unions and others. The volume of lip service is deafening but positive action is rare. Few, if any, countries can claim to have adequate facilities. Practically all the medical and para-medical resources of countries are devoted to therapeutic medicine and surgery. The less glamorous branches of preventive, occupational and rehabilitative medicine tend to be the Cinderellas.

The needs are obvious. Serious problems beset anyone who becomes disabled from his or her usual job and these can and do affect both the individual and the family. There is also an effect upon the economy of the organisation and of the nation. Most modern societies operate a system that involves each person at work supporting about two who do not, for example the young, the old, the unfit and so on. To allow large numbers who could return to work, and thus generate wealth, to become dependent, is clearly foolish in economic terms. It is also wrong to let people who could work again to become vegetables for want of the stimulus of work. They run a serious risk of becoming isolated from society—human rejects—for want of the dignity which independence and self-sufficiency allows.

Retraining and resettlement are also required on an ever increasing scale as a result of the inevitable technological changes found in modern industrial society. This is not of course primarily a medical problem, but lack of provision to cope with these needs also causes isolation and depression and thus ill-health amongst those who are made redundant. There are, however, obvious benefits to be gained by encouraging people who have to be resettled for medical reasons to retrain alongside the healthy. The psychological aspects of rehabilitation and resettlement are every bit as important as the more obvious physical ones. Those responsible for retraining courses are well aware of the need for careful selection of those to be taught. A high rate of successful job placement is essential since this affects the morale of those in training. Someone who is unable to continue in his chosen career, whether by reason of ill-health or the obsolescence of his skills, is easily demoralised. The fear that he may not be able to find a job even after retraining is enough to induce an apathetic lethargy which is difficult to overcome. People disabled by psychiatric disorders tend as a group to be the most difficult to resettle, and most retraining courses prefer to limit their intake of such people to about a quarter of the total. The prejudice of prospective employers, already described with reference to diabetes and other slowly progressive diseases, is even more of a barrier to the satisfactory resettlement of people with mental disorders. There was one case of an anti-social relapse in such a patient which readily produced the attitude of 'once bitten twice shy', but this is uncommon and such tales are more often heard at second or third hand. The revolution in treatment of psychiatric disorders in the past decade has greatly improved the outlook for such people.

What then accounts for the unsatisfactory state of affairs we describe? Criticism of the attitudes of most doctors towards rehabilitation can be heard in countries all over the world, although to be fair, similar comments are also made about others who should also do more. Allegations that a majority of doctors take a disinterested or tolerantly passive role are probably true. One reason seems to be that most doctors view their patients as people only to be helped or cured of illness or injury, rather than as people who must be helped to return to an economically active role in life and to the independence which this brings.

It has often been stated that rehabilitation should begin at the onset of an illness or injury. Only seldom is this actually achieved, except in special units such as the one for spinal injury at Stoke Mandeville. This attitude is one of the reasons for its world-wide reputation. With some notable exceptions, hospital doctors are often the worst offenders; only when the medical condition has improved and a date for discharge is being considered may a thought be given to a return to work. Even then, it is often the social worker or the patient himself who raises the matter. A few hospitals have recognised this deficiency and employ a resettlement officer who acts as a link between the medical staff and the management of industry and commerce in the area. This has been found to be a most useful and effective arrangement.

The doctor or nurse in industry can also be of great help in rehabilitation and resettlement since they should know the extent to which existing jobs can be modified or alternative work provided on a temporary or permanent basis. They, or someone from the personnel department such as a welfare officer, should also be aware of the schemes in the community to help the disabled, and of the voluntary groups which provide assistance for special conditions such as alcoholism, diabetes, blindness, deafness and multiple sclerosis.

The aim should always be to get the person back to his *normal* job as soon as possible. Where the impairment prevents this, it may be possible to allow him to return to a *modified* job amongst his own colleagues. A person's own workmates are much more likely to allow him to fit back into the group than are a group of strangers. All work has an important social element, and a return to work after incapacity involves taking up a social position in the working group as much as it does restarting the work itself.

Effective medical advice to managers must be couched in terms that are both constructive and realistic. The practice of rehabilitation and resettlement requires that everyone concerned with the individual should

Disease, Impairment and Disability

communicate effectively with each other as well as with the person concerned. Unfortunately this state of affairs does not always occur. The use of a phrase such as 'fit for light work' causes as much, if not more, trouble for employer and employee alike as any other single medical pronouncement. The interpretation of light work by the patient may range from doing virtually nothing, to lifting half his usual load. To the employer it may mean that he must create a job that does not really need doing, but for which he is expected to pay a full wage. When the person has been given a recommendation for 'permanent light work', it can even be tantamount to permanent unemployment. Where there is a doctor or nurse at work they should be able to say what jobs the man could or could not do. Where there is not, the employer could reasonably invite the doctor who issued the certificate to attempt to spell out just what he considers his patient could or could not do, in physical and psychological terms. The answers might sometimes be illuminating!

Rehabilitation workshops, since they require a very large population employed to generate enough cases, are usually provided by the State. The facilities provided for rehabilitation can seldom cope with the number who would benefit. In some countries rehabilitation centres may be provided by the combined resources of an industry, coalmining is perhaps the most obvious example, and sometimes by large individual concerns such as motor vehicle manufacturers. Although many of these centres are well planned and really do speed recovery to maximum function, there are others that are little more than sheltered workshops in which no active rehabilitation is practised.

Employing the disabled

In times of severe shortages of manpower everyone able to work is encouraged to do so. Conversely during periods with substantial unemployment people with impairments find it most difficult to get jobs. In Britain, for example, a system of registration, which is entirely voluntary, was established in 1944. Anyone with an impairment which substantially handicaps his employment opportunities in relation to his age, experience or qualifications, and whose condition is likely to last for not less than one year, can apply for registration as a disabled person. The legislation also set up the 'quota' system under which employers of twenty or more people are required to have at least 3 per cent of their staff formed by registered disabled persons. This law was primarily designed to assist disabled people to obtain work. As a consequence, a person employed as disabled remains part of the quota even if his medical condition clears up.

It has been demonstrated that about one-fifth of registered men in employment have impairments that have either recovered or have had no significant effect upon the type of work for which they were normally suited. Such people do not require the assistance that registration can give and they block places in the quota that might be more usefully filled by others. The main problem, however, is that the voluntary system, coupled with the innate reluctance of many to accept registration as 'disabled', means that the system is not as effective as it could be. There is good reason to suppose that only about one disabled person in three actually registers. Surveys have shown that there are far more disabled people in and out of employment than the official figures suggest. One of these, covering over 12,000 men in eight different types of factory, showed that 11 per cent could have been registered although only one-third of them had actually done so. Registration was neither associated with severity of disability, nor did it seem to be related either to the medical condition or to the type of work. It is of importance to note that the proportion of disabled is as high as this, and that it involves almost one man in four over the age of fifty-five. This raises difficulty when implementing productivity agreements which require—as they usually do—greater flexibility of work. There are many people with impairments who are able to do their usual job without limitation, but who become disabled if required to do other work. Productivity agreements between managements and unions do not usually have special clauses designed to protect such individuals. It would, moreover, be quite wrong for anyone to think that a disabled person—whether registered or not—is necessarily going to be a burden to the organisation; his sickness record and his performance may well prove better than the normal. Disabled people are often very good at their jobs. This fact should be much more widely appreciated than it is at present.

Personal adjustment to impairment

In the course of our work in industry we have been struck by the lack of consistent relationship between objective measures of health and absence attributed to sickness (p. 40). This is particularly true when people with disabling medical conditions are considered. There is little doubt that the attitude of the individual towards himself, his medical condition and his work, is a great deal more important than the results obtained from a physical examination. The people who adapt best to a physical limitation are generally those in whom the condition is static and has been present for some time. A condition which commenced in childhood, particularly as the result of injury, seems to be the easiest to adapt to; the speed with which children adapt to change of circumstances is well-known. There is however no evidence to suggest that any particular type of condition is likely to cause poor adaptation. We

Disease, Impairment and Disability

have seen men with what on objective medical terms are minor conditions, who seem to bear a grudge against life, society and their medical condition in particular, who spend much of their time off sick or traipsing from one hospital or specialist to another. They form but a small minority of the total number of people with impairments (10–15 per cent), but their impact upon their managers, colleagues, friends and relations is out of all proportion to their numbers. They also usually show other signs of maladjustment to life which may be detectable at an extended pre-employment interview.

The value of pre-employment medical examinations, as usually performed, is strictly limited. The medical condition and its consequent impairment of function can be assessed. Psychiatric disorders, however, can be difficult to diagnose because of the constraints and inhibitions imposed by this sort of examination in which the relationship is very different from the usual one between doctor and patient (p. 29). Assessment of disability requires more evidence of adaptability than is usually available, except for the most clear-cut of conditions. The prediction of absence attributed to sickness can be little more than an inspired guess, unless the condition is one that has a particularly poor outlook. There is little doubt that a well-adapted disabled employee can often attend more regularly and work at least as well, if not better, than his nominally healthy colleague.

Finally a comment about pension funds. There are a disturbingly large number of cases reported every year to the voluntary societies for diabetes, epileptics and the like, of refusal to employ 'because our pension fund will not permit it'. This is probably often untrue, although difficult to prove without access to the pension fund trust deed. Moderately large organisations which operate their own pension schemes sometimes seem to be reluctant to carry their share of the disabled in the community. The funds both of very large organisations and of smaller firms that use insurance companies to provide pensions, usually have an actuarially calculated system to permit the employment of people currently fit to do the job for which they are recruited, even though the long-term outlook may be poor. Thus the very largest and the smaller organisations are usually less restrictive than the moderately large. Some doctors who undertake recruitment examinations seem to take it upon themselves to protect the firm's pension fund even if they have not been asked to do so. We would suggest that doctors doing such examinations ought to be informed about the view of the firm's pension authorities, for in our experience they are seldom told anything at all about this. It is clearly in the interests of all that disabled people should be given an equal opportunity of *suitable* employment as the rest of the workforce; unfortunately they are often the victims of prejudice. This is manifestly wrong.

Further reading

The Disabled in Society, P. Townsend, Greater London Association for the Disabled, London (1967).

Handicapped and Impaired in Great Britain, Amelia I. Harris, HMSO, London (1971).

Work and Housing of Impaired Persons in Great Britain, Judith R. Buckle, HMSO, London (1971).

Rehabilitation, Report of a Sub-Committee of the Standing Medical Advisory Committee of the Department of Health and Social Security, Welsh Office, and the Central Health Services Council, HMSO, London (1972).

CHAPTER 7

Shift Work and Health

Shift work of one sort or another has been the normal pattern for some people throughout history—for example, nurses, sailors, bakers and policemen. For these occupations shift work was clearly necessary to satisfy the needs of society. Industrialisation had a need for shift work from its earliest beginnings where manufacture required continuous operation as in iron and glass-making, and in modern times in the production of power sources such as gas, electricity and the refining of oil. Here, too, workpeople appreciate the technological necessity of continuous operation. Today, however, the rapid growth of shift working is largely for economic reasons. The very high capital cost of complex machinery—from computers to vehicle assembly lines, where this year's model is obsolescent by the time it has been produced—requires maximum utilisation in order to earn profits in the limited time available. This is a justification for shift and night work that some people find less acceptable by comparison with the social and technological reasons in the past.

The subject is introduced in this way because it provides an explanation for the increasing resistance that some organisations are now finding to the introduction of shift working. There is little doubt that people in towns where continuous process or heavy industry have been established for many years are a great deal less resistant to the idea of shift work than people in places where there is no such tradition. The scale of shift work in modern industrialised countries is now considerable. For example, the number of shift workers employed in manufacturing industry in the United Kingdom increased by at least 50 per cent in the course of the 1960s. It has been estimated that about one employee in three in the United Kingdom is required to work at times outside normal day work hours for a substantial part of his or her working life. This estimate includes people employed in service industries where shift work is extremely common.

Types of shift work

Since any one organisation or factory tends to operate only one type of shift system, many people think of shift work as being the type that they have met. This alone can lead to misunderstandings since the term 'shift work' covers a wide variety of arrangements of working hours, and it is essential to appreciate this. Shift work includes any irregular arrangement of working hours which involves work outside the times of usual day work: 8.00 a.m.-5.00 p.m. or thereabouts. It therefore includes work at night and also the early evening or 'twilight' shift.

A full classification of shift systems can be complicated, but it is worth considering examples of the two main groups: those involving work around the clock for twenty-four hours each day and those involving less than that. The former is usually covered by three shift crews, each working for eight hours, and this is therefore called the *three-shift system*. When the factory only operates for five days each week this system only requires three crews (discontinuous three shift), but when the process runs for seven days each week a fourth crew is required for such a continuous three-shift system. In both arrangements it is usual for the crews to 'rotate' by taking turns at the morning, afternoon and night shift. There are, however, other ways of manning a process around the clock. One is to operate twelve-hour shifts, and the other is to have one crew on *permanent night work* with the remaining hours covered by *double-day shifts,* one starting at about 6.00 a.m. and the second at about 2.00 p.m.

Shift systems covering less than twenty-four hours can differ even more widely. Apart from the system of double-day shifts which can provide cover for sixteen hours, there is the *twilight shift* of about four to six hours at the end of the normal working day which has been found to attract married women who are able to go out to work when their husbands get home and can look after the children. The *split-shift system* is most often found in the catering and transport industries since they have peak work-load times separated by several hours. This usually involves starting early, finishing late, but having time off during the day. As can be imagined this arrangement is only practicable when people live fairly close to their place of work. There are also the complicated arrangements found in some radio and television workers in which the length of the working day and the time of starting can vary in a most confusing way. Finally, many shift workers in vehicle-manufacturing do alternate day and night shifts with the intervals between given over to maintenance work on the production line.

All these arrangements present their own unique problems of adjustment to the men or women who work them, and these problems often

Shift Work and Health

affect their families as well. Such a wide variety of systems should never be thought of as a single entity and therefore the term 'shift work' should not be used without qualification of its type. Failure to do so can cause difficulties through over-simplification.

Even one system such as the three-shift continuous variety providing round the clock cover for 365 days each year can be arranged in several quite different ways. Although it has been usual for shift crews to rotate by changing the shift at weekly intervals, some organisations do this at fortnightly or occasionally even longer intervals. Others, however, make the change more frequently and this system of rapid rotation every two or three days has become quite popular in the past few years. There is the so-called *continental* system ($2\times2\times3$) which takes four weeks to complete a full cycle, and the *metropolitan* system ($2\times2\times2$) which takes eight weeks. Even with a given frequency of rotation, the order in which morning, afternoon and night shifts rotate with rest days can differ by any permutation of these four variables. The actual hours of shift change may also vary and the night shift in some factories may be one or two hours longer than either of the other two. The so-called *traditional* system of three shift weekly rotation of mornings, afternoons, nights and rest days in that order involves the very unpopular 'dead fortnight'. This is named because a week on the socially useless afternoon shift (2.00 p.m.-10.00 p.m.) is immediately followed by a second week on the maritally unpopular night shift. This is often a bitter source of complaint for the men who work this system. It is worth adding, however, that a remarkable characteristic of shift workers is their conservatism and reluctance to change to a different system from the one to which they have become accustomed.

One further problem should be mentioned briefly in relation to shift work, and this is the two-job man. He is most commonly seen in industry as a shift worker. Holding down two jobs may lead to all sorts of problems involving shift rota acceptability to say nothing of health problems resulting from excessive hours.

These, then, are points to be borne in mind before considering the consequences of shift and night work on health. With so many workers employed on irregular hours of work for so many years, it may seem that all questions of health should long since have been resolved, or at least agreed. Nothing could be further from the truth. Apart from some studies at the time of the First World War the scientific study of this subject has been almost completely neglected until quite recently. The trade union movements in many countries, however, have been consistently opposed to work at night except where it is clearly unavoidable for social or technological reasons. At one of the first international trade union conferences nearly a century ago, the movement declared its complete opposition to the use of night work, and many unions still have this as their declared policy, even though its implementation may have proved difficult.

Shift work is said by those who oppose it to be anti-social, destructive of family life, unduly fatiguing and, on some occasions, damaging to health. All these assertions have been cited in the past and are still being used today to limit the introduction of shift work. When shift working is inevitable they are used to justify the payment of a substantial shift allowance as compensation. Much of the evidence for these assertions, particularly in regard to health, is of dubious validity.

The health of shift workers

The whole problem is heavily coloured by emotion and, even worse, by rationalisation. As a corollary of 'a little of what you fancy does you good', there are those who argue that as work at night is against the laws of nature and is unpopular, it must therefore be harmful. When to this is added the view that it is morally or socially damaging, the seeds of prejudice and obstinacy are sown. On the other side of the coin, some employers appear convinced that since they and others can easily recruit volunteers for shift work, and since shift workers seem to live to a ripe old age, the alleged health problems are grossly exaggerated and can in any case be settled easily by a slightly larger financial incentive. We would subscribe to neither view. The truth we believe lies somewhere in between, even though the evidence shows that shift work and night work has no adverse effect on the health of the overwhelming majority of those who have worked these systems for several or indeed for many years.

Practically all the medical evidence which is quoted to show that shift work damages health is based upon subjective opinion and not on objective fact. Perhaps the view most widely held by doctors and laymen alike is that shift work causes peptic ulcers. The evidence of several carefully conducted studies is now substantially against this. Many of the arguments which have been produced to show the ill-effects of shift work are based upon direct questioning of workpeople about the symptoms and difficulties which they attribute to shift work. To be valid these questionnaires must be 'blind' in that both shift *and* day workers should be questioned and that neither group should know that shift work is under investigation. An excellent study of nearly two thousand workpeople in the Netherlands a few years ago did just this and demonstrated no medical complaints related to shift work. The only difference between the groups studied was that shift workers answered a question about dropping off to sleep on getting home from work in the affirmative significantly more often than day workers. Since two of the three shifts ended at times when many of the day work population are still abed, this is not altogether surprising!

Another source of evidence comes from analysis of sickness absence records. The limitations of this as a measure of ill-health are discussed on p. 36. However, the results of such studies are perhaps surprising. Virtually all have shown fewer spells of sickness absence and fewer days of absence amongst shift workers. These reports have come from the United States, the Netherlands, West Germany, Norway and the United Kingdom. Earlier studies had been made amongst three-shift continuous process workers, but in the last few years evidence has also come from other types of shift work in other types of industry. Some of these studies also analysed non-medical absences and showed that these, too, were less frequent amongst shift workers. A small minority of firms studied did not follow this general trend, and here the evidence suggested that this might be attributed to longer hours of work of some shift workers and also to unusually high shift differentials. Nevertheless, the general trend of less absence attributed to sickness and also to other reasons must not be taken to imply that shift workers are abnormally resistant to certain diseases. They seem to have all the usual ailments in the usual proportions, but simply to absent themselves from work less frequently than their colleagues on day work. It has been suggested that their motivation to attend is stronger, and where they have shift handover arrangements, the unexpected absence of one relief can greatly inconvenience the man to be relieved. Some shift workers also express a preference for work outside ordinary day hours because there are 'fewer bosses about'. The reasons are complicated and clearly need more study before it would be safe to generalise.

Most important of all medical aspects of shift work, however, is any effect it may have upon death rates. Some shift workers believe that the hours they work can shorten their life. There is no evidence whatsoever to support this view and indeed two studies, one in Norway and the other on a larger scale in the United Kingdom, have shown that death rates are the same amongst shift and day workers in similar industries and occupations.

It should not be concluded, however, that shift work presents no medical problems at all because of the lack of convincing evidence that it can damage health. For obvious technical reasons all research studies have been done on groups of shift workers who have worked on shifts for months and sometimes for years. This means that the findings are based upon populations of 'survivors', that is, no account has been taken of those who, for various reasons, ceased to do shift work after a short period on it. There is reason to believe that something like 10 per cent of any group of people will find that they cannot adapt themselves to shift work. The studies on health mentioned above excluded, as far as possible, any man who had been taken off shift work for reasons of health because their inclusion among the day worker group could distort the results. We shall discuss below the possible reasons for this failure to adapt to shift work, but at this point we would simply observe that provided shift working for the individual remains mainly voluntary, there should be no serious problems of ill health due to the arrangement of working hours.

Circadian rhythms

The study of inherent biological rhythms, known as circadian rhythms because they vary over a twenty-four-hour period, has received a great deal of attention in the past ten years. Man, and indeed all living matter, shows many such rhythmic fluctuations in terms of biochemistry, physiology and behaviour. We are all familiar with these in terms of sleep, wakefulness, digestion, elimination and so on. Most people know, for example, that body temperature is slightly higher in the early evening than it is in the morning. It would now be true to say that virtually every biochemical and physiological variable that one might choose to measure would show a twenty-four-hour rhythmic variation which becomes established in infancy and persists throughout life. These variations are not merely of academic interest since they account in part for the real problems experienced when one flies rapidly across several time zones, or when a person first goes on night work. Although the rhythm of body temperature can adapt within about three or four days after a flight from England to New Zealand, other rhythms, such as those of renal excretion of certain salts, may take up to a fortnight to adjust to the new time. The night worker's problem is more complicated in that the environmental day–night pattern remains unchanged as does the bustle of life at home. A few years ago the Medical Research Council suggested that the frequency of shift change should be at intervals of more than a week to allow for better biochemical adaptation, but this would only be effective if people did not take days off in the middle. Such modification of body rhythms that a night worker can achieve are all lost after forty-eight hours if he reverts to an ordinary day time life at the weekend. Few industrial workers would accept a proposal that they should regularly work for two weeks without a day off and, if they were to do so, they would probably run into problems of fatigue. Some research workers have suggested that the rapidly rotating systems are preferable to the traditional weekly change since biochemical adaptation takes even longer than this and any disadvantage of this sort is offset by the attraction of only working for a couple of nights at a time.

Circadian rhythms and working efficiency

Managers often enquire whether day or night work is the more efficient. Most studies in factories have shown little difference or a slight advantage for day work. Recent studies by ergonomists on carefully controlled laboratory conditions have shown that the speed and accuracy of various tasks run in parallel with the daily fluctuations in body temperature. That is to say performance is at its best in the late afternoon or early evening and at its worst in the early hours of the morning. We must however warn any enthusiast that abnormally high temperatures do not further improve performance! In real life situations the effects of fatigue tend to outweigh these fluctuations. There is also good laboratory evidence to support the well-known phenomenon of the dip in alertness and efficiency after a good meal. This can last for about an hour after a three-course meal. Accident studies of day and night workers have usually shown a lower rate at night, but this is probably due to the somewhat lower tempo of work and the more relaxed atmosphere usually found at that time.

Social and family effects of shiftwork

The social consequences of shiftwork, and of nightwork in particular, provide most of the problems of adjustment both for the worker and also for his family. Shift work seldom attracts the young man, particularly whilst he is still unmarried. The afternoon shift (usually from 2.00 p.m.-10.00 p.m.), whether as part of a three-shift system or of a double-day shift, seriously inhibits participation in social activities such as games, and often, even more important, meeting girls ('bird-watching is out' as one young shift worker explained). Some studies by questionnaire have shown that this shift is slightly more unpopular than the night shift. By the time the worker gets off duty most of the bars are closed. For this reason it is not uncommon to find the night shift crew coming in half-an-hour early to allow their colleagues time for a drink on the way home.

Night shifts are also unpopular with married men, often because their wives dislike being left alone. Suggestions that night work is a contributory factor in the break-up of marriages have not been confirmed because as many men and wives state that their marriage is improved by shift as claim that it is damaged. Indeed, it is true to add that there are some less happy marriages that have only survived because of shift work. The complaint most frequently voiced by wives is the difficulty that rotating shift work brings in parental relationships. The wife may be forced to take up a paternal disciplinary role not to her liking. During school holidays there are the additional problems of keeping the children quiet while father tries to sleep in the daytime. The morning shift (usually 6.00 a.m.-2.00 p.m.) although the most popular of the three, involves rising early and can thus curtail social activities in the evenings.

In summary, the afternoon shift interferes with the parental role and with the social life of the worker, while the night shift tends to affect the marital role. Any effects upon the children in terms of delinquency have not, as far as we are aware, been studied. Shift work for some has great compensation. Those with hobbies such as gardening or fishing are able to enjoy them in daylight throughout the year. Travel in the peak rush hours is avoided and shopping expeditions are possible when most people are at work. Occasionally, the extra money gained by shift work may be mopped up by the cost of providing independent transport because public services may not provide transport at shift-change hours. This can lead to discontent, and perhaps to stress, when the expected gains of doing shift are dissipated in this way. However, it is the experience of all who have studied this problem that a substantial proportion of shift workers view the prospect of changing to straight day work with little enthusiasm. Even though they may originally have taken up shift work because of the financial inducement, they remain on it because they learn to prefer it.

Conclusions

The marked preference of shift workers for the arrangement they are accustomed to has already been mentioned. Thus change from one shift system to another is not common. There have been a few reports in the scientific literature in recent years about changes from the 'traditional' weekly three-shift to the rapidly rotating systems. All have emphasised the social advantages of the change with fewer nights away from home at a time and the disappearance of the 'dead fortnight'. Only one such change involved a sufficiently large sample to allow a worthwhile study of absence. This showed, to the authors' surprise, a rise in sickness absence but a fall in non-medical absence. The only conclusion one may draw from a single factory is that the demonstrated increase in social acceptability of the new system was unrelated to absence attributed to sickness. More such studies are clearly necessary. The problem of how to cover a night shift most satisfactorily has yet to be agreed as also has the problem of agreeing the ideal shift system. Acting on the principle that unpleasant work and financial compensation should be equally shared, the unions tend to prefer rotating shift systems. This is not necessarily the best arrangement since most working communities contain enough people who prefer to take on permanent night work. Studies of such men have usually shown that they are a settled and contented group. We would suggest that to cover a twenty-four-hour period, particularly with discontinuous five-day operation, it may prove most satisfactory to have a double-day shift and a permanent night crew.

We would accept that shift and night work does present problems both to the employer and to the worker. The problems may be medical, physiological or psycho-social. Of these, the psycho-social are the most important. Overall, however, the problems are not so severe as some have tried to make out. As long as the individual worker can maintain a degree of personal choice, and can in the last resort opt out of shift work if he cannot adapt to it, the problems should never be too serious.

Further reading

Shift Work, Paul E. Mott, Floyd C. Mann, Quin McLoughlin & Donald P. Warwick, University of Michigan Press (1965).

Managing Shiftwork, Robert Sergean, Gower Press, London (1971).

Hours of Work, Overtime and Shiftworking, National Board for Prices and Incomes, Report No. 161, HMSO, London (1970).

Biological Rhythms and Human Performance, W. P. C. Colquhoun (ed.), Academic Press, London (1971).

CHAPTER 8

Occupational Hazards and How to Deal With Them

Occupational disease

Occupational disease has been defined as disease which arises out of or in the course of a person's employment. The causal or associative connection may be easily recognised in certain instances while in others it may present very great difficulties. Legal problems can arise where difficulties appear when defining illness or disease as an occupational one or not.

Measuring health and disease

In any community or group, the health or ill-health of the group or of the member of the group can either be guessed or measured. In the face of an influenza epidemic, which may affect as many as half of the people in a time span of a few weeks, it would be an insensitive person who did not spot the rise in disease. Occupational diseases, however, often tend to be insidious in onset and the result of long-term minor repeat exposures to a hazard. It may only be by measuring health and disease that such problems can be brought to light. The disease may be unusual—in which case it may be recognised (p. 69). On the other hand, there may simply be more of a certain generally recognised and apparently ordinary disease in an occupational group.

Recognising and diagnosing occupational disease

When the time interval between cause and effect is short there is usually little or no difficulty in recognising the causal relationship between event(s) and diseases or injuries. However, many occupational diseases take time to develop and when a person with a disease is seen by a doctor who asks himself the question 'could this disease be related to the person's occupation?' the answer may be anything but simple. The relationship of the disease to occupation may be either that of *cause* or that of *association*. The environment and the sickness may be causally related or the two may, by chance, be found together. How can this relationship be sorted out? Sir Austin Bradford Hill in his book on the principles of medical statistics listed a number of features which should be examined when trying to determine the nature of any relationship, whether causal or merely associative:

1. The strength of the association

Do the observed features of the disease occur in everyone, in a large number of people, or only in a few of the people working with substance X and in that particular part of the factory only?

2. Consistency

Have the observed features of the disease been found in only one factory by only one group of observers, or have a number of different observers of different people and groups found the same or similar disease patterns in different places, circumstances and time?

3. Specificity

Is the disease specific to this group of workpeople and to particular sites and types of disease not found in other people? If so, then there is a strong probability that the relationship with occupation is causal.

4. The relationship in time

The 'chicken and the egg problem' is a good example—which came first? The time relationship of suspected occupational disease in a group of people may give useful information about cause or association.

5. The biological gradient

A dose-response relationship, for example, more heavily exposed people having worse or more rapidly appearing effects in the way of disease is always in favour of cause rather than of association.

6. Biological plausibility and coherence

Any theory advanced about the causality must be tenable in the light of existing knowledge. If the answer from existing knowledge is 'don't know' the theory may be acceptable. However, if the answer is clearly that 'this cannot be so', then a new explanation will be required. However, seemingly implausible explanations which cannot be disproved *may* be right and should not be discarded too readily.

7. Experimental proof

In some instances it may be possible to prove causal connections in this way. For example, if in a group of workpeople some were regularly suffering from a form of dermatitis and then a substitution was made of one substance and from this time onwards no more cases of dermatitis arose, a cause and effect relationship may be inferred.

8. Argument by analogy

In industry an example might be that in looking at a problem of suspected occupational disease it was known that a very similar and closely related chemical compound caused a similar illness in workers elsewhere. Knowing this, is it likely that there is any other way of explaining this outbreak of disease, and is the likely explanation one of cause and effect?

Injuries at work and occupational diseases

A light-hearted but basically serious description of the difference between occupational injuries and occupational diseases, is that injuries get you at once whereas diseases get you eventually. The time-scale is quite different in most instances. Occupational injuries often called 'accidents', though they are far from accidental, account for many more people being off work on any given day than do occupational diseases. The relative frequency of occupational injuries and diseases for the two sexes are shown (see figures 2 and 3).

The causes of injuries are multiple, and usually act within a short time of the injury occurring. A scientist working in a research laboratory went out to lunch, leaving an experiment running. Owing to a failure in a thermostat, an oil heating bath became too hot, caught fire and set fire to the bench. A fellow scientist was burned trying to put the fire out. The causes of this incident, which led to the injury, were multiple and included going away and leaving an experiment running, equipment failure, flammable materials on flammable benches and lack of knowledge of fire-fighting amongst other scientists. All of these causes contributed to the eventual outcome and all were operating within a short time of the injury. Had some of the causes been recognised and dealt with at an earlier stage, the eventual burn injury may have been prevented. The damage to buildings and equipment would also have been prevented.

Occupational diseases usually have a much longer time scale between cause or causes and effects. A foundry worker made a casting material out of a silica-containing powder. He also helped at times to remove

Occupational Hazards and How to Deal with Them 65

FIGURE 2: *Days of incapacity in United Kingdom insured population.*

MEN

attributed to work
(7 per cent)

WOMEN

attributed to work
(3 per cent)

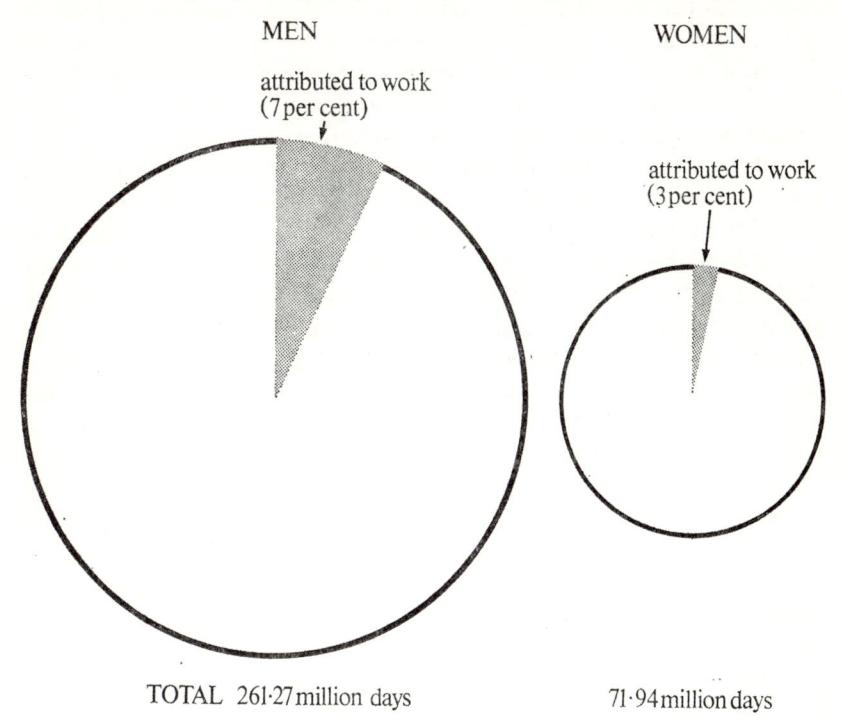

TOTAL 261·27 million days

71·94 million days

FIGURE 3: *Days of incapacity attributed to work in United Kingdom insured population.*

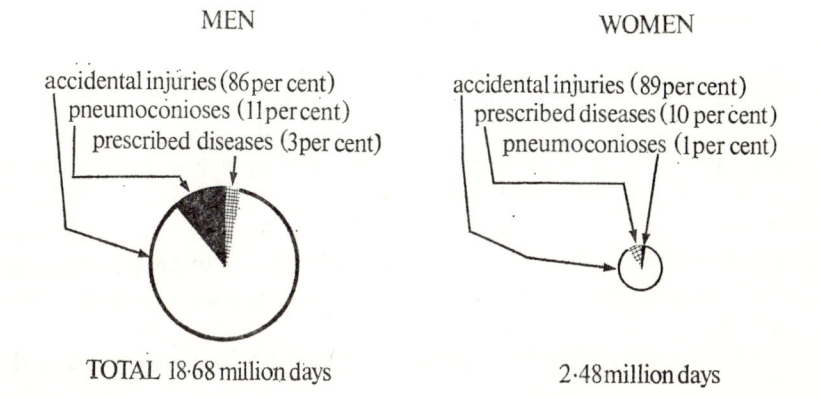

MEN

accidental injuries (86 per cent)
pneumoconioses (11 per cent)
prescribed diseases (3 per cent)

WOMEN

accidental injuries (89 per cent)
prescribed diseases (10 per cent)
pneumoconioses (1 per cent)

TOTAL 18·68 million days

2·48 million days

the casting material from the finished casts. In both operations he was exposed to the inhalation of small amounts of silica-containing dusts. A mass X-ray of his lungs twenty-one years after he was first employed showed evidence of a fibrotic occupational lung disease called silicosis due to the inhalation of these dusts although the man had not yet noticed any symptoms nor did he feel in any way unwell.

If this man had been asked if he felt anything wrong with him after five years of work his reply would probably have been negative. Similarly, his foreman might have believed that the dust was harmless or replied that there was so little of it that no harm could result. In any case, he had been there ten years and was perfectly well so that the process was therefore quite safe. Such 'reasoning' occurs all the time in connection with the identification and prevention of occupational disease. In some exposures to silica the disease became crippling within ten years. This led to the saying 'join the Navy and see the world; become a sandblaster and see the next'.

The biological insults which result in eventual disease are often not recognised as being causal because they are small, regular, part of doing the job, always present, not connected in people's minds with the eventual outcome and frequently dismissed from the mind even when reason can show a connection, for example, wearing or not wearing respiratory protection when working with dust or fumes or, in a non-occupational context, people persisting in smoking cigarettes even when they are aware of the risks, lung cancer, chronic bronchitis or heart attacks.

Occupational diseases tend, therefore, to be more insidious in origin than occupational injuries and to require greater levels of awareness in the observers to detect possible hazards. There are, of course, some occupational diseases which can appear quickly, for example, illness following gassing at work, allergies from substances met at work and the effects of excessive amounts of radiation or explosive noise.

What to tell workpeople about hazards

It is our belief that every person should be told the truth about the dangers of every job which he has to do and about the hazards of every substance which he may be required to handle or to contact. To conceal hazards in the short term may be successful but deceitful; in the long term, the truth will out and trust cannot be regained. Men and women who are not told of the dangers in their working environment are deprived of information which they should use to protect themselves and become, as it were, children in an unsafe environment. This is wrong.

From time to time the argument is put forth by timorous—and dare

Occupational Hazards and How to Deal with Them

we allege inadequate—managers that there will be strikes, demands for money, bad union relations and so on, if the truth is told. This should not be so. We have yet to meet a case where full explanation of the hazard(s), however alarming these may at first appear, coupled with education, is not followed by a lessening of tension, by trade union co-operation and by the realisation that health is too important a matter to bicker about. Education will inform everyone about what management is doing to control the hazard and what each person should do to co-operate in the safety programme and work safely for himself and others. If people are treated as adults and as responsible people they will tend to behave in this way.

When good management replaces bad, there may be an interim period while unions and people on the shop floor have to learn to trust each other. This will be a testing time and the way in which hazards are explained is of the greatest importance.

In one chemical plant, for instance, one of the units began to cause trouble and a toxic fume was released from leaking joints. The trouble was spotted and working commenced before the workpeople had made any complaint. It was suggested that the plant manager should go to see the situation for himself but he refused because, he said, the men would see him and immediately realise that something serious was amiss.

How substances can cause harmful effects

In order to produce biological damage, a substance must gain entry to the body or be in contact with it. The main routes of entry of substances into the body are by

1. *Inhalation.* Breathing in vapour, gas, mist, spray, fumes or dust is by far the commonest and most important route of entry of harmful or poisonous substances in any industrial or work situation.
2. *Ingestion.* Swallowing is the commonest route of entry in home poisoning, and may occasionally occur at work. It is, however, infrequent as a route for industrial poisoning except through eating or smoking without first washing the hands.
3. *Through the intact skin.* The skin does not act as a barrier to some substances, particularly solvents, and absorption of these substances can follow skin contact. General poisoning may thus result from local contact.
4. *Injection.* This is an unusual route to entry in a work situation.

Contact with chemical substances can cause harm by:

1. *Burning.* For example, acids and alkalis. The eyes are especially susceptible to damage in this way. Alkali burns are usually worse than acid burns.

2. Local damage. The skin is easily damaged by a wide range of chemicals following skin contact. A number of mechanisms are involved, one of which is *de-fatting* of the skin from contact with fat solvents such as hydrocarbons and solvents. Severe dermatitis can follow from these causes.

3. Allergic reactions. The skin is easily affected in this way by a variety of substances. Other allergic reactions such as asthmatic attacks are also possible.

4. Absorption leading to general poisoning. This has been discussed above.

The principles of dealing with any occupational hazard

Three words can be used to outline briefly the steps which may be taken in order to prevent occupational hazards from becoming occupational diseases: identification, evaluation and control.

Identification

Hazards which are unrecognised will remain uncontrolled. Stated in this way it may appear as blindingly obvious. In practice, it is a reality which is only too painfully engraved on the consciousness of everyone who has ever worked in the field of prevention of occupational disease. Two examples which relate to non-recognition of occupational disease will illustrate this. In the first, the engine-room staff of a large ship cleaned the ship's boilers at sea. Many of them became ill. They had inhaled vanadium pentoxide which was contained in the boiler-soot scale of the oil-fired boiler. They did not recognise the hazard of vanadium being concentrated from mineral oil in the boiler-soot, nor did they have any idea why they were ill. The second example involves an office junior who felt nauseated and sick. She was using a duplicating machine from which the solvent evaporated into the poorly ventilated work room. Inhalation of solvent was the cause of her illness.

Similar examples could easily be given in relation to injury and fire hazards.

The worst way to identify or indeed to monitor any situation where an occupational hazard may arise or has arisen is to make use of what could be described as the human canary approach. When the human canary falls off his perch something must be wrong! In a similar way early wing flap or instability on the perch or not singing normally, or some other symptom, could indicate early toxic effects. None of these methods will do. We once met a brewery man whose job it was to clean out the fermentation vats. These were always blown through with fresh air to remove the carbon dioxide formed during fermentation. He was provided with a canary to monitor the air before he entered. When

Occupational Hazards and How to Deal with Them 69

he was asked how this arrangement worked he replied that he had stopped using canaries because they often died! We swear to the truth of this story.

It is, however, perfectly reasonable to use personal dosimeters or biological monitoring as a back-up method of control to ensure that the methods of environmental control are working properly, but environmental control must have prior attention. For example, people who may be exposed to radiation should always wear a film badge or dosimeter which will show the amounts of radiation to which they have been exposed. Screening and monitoring of radiation must, however, have precedence as a preventive method. In other situations, such as exposure to certain chemicals, use can be made of blood and urinary levels of these chemicals or of their breakdown products in order to detect early absorption or higher than normal excretion rates. However, even before environmental control the question should be asked, 'why do it at all?'.

Suspecting a hazard

A high index of suspicion is always better than a low one. Identification of hazards will be more frequent if a high index of suspicion is maintained. This can be done by:
1. Observation of the environment.
2. Observation of workpeople.
3. Complaints from workpeople.
4. A high level of sickness in a group of workpeople.
5. A high level of sickness in an individual.
6. Unusual illness in one or many people.
7. Good records of illnesses and deaths used to compare known figures with expected figures.

Evaluation

Evaluation of a hazard requires, among other things, a complete knowledge of how the substance(s) will be handled, how the process(es) operate and what exposure(s) may be likely. It may be necessary to review engineering design, to assess hazard potential (which may or may not be known in detail), to measure the concentration of dusts, liquids, mists, sprays, gases or fumes which may be present and to compare these with known threshold limit values (see below).

Evaluation must include the total work area and the total range of substances handled: this will include intermediate products as well as finished products. Subsidiary processes must all be included in the evaluation because if any aspect is left out, the opportunity to prevent danger may be missed.

Occupational hygienists (sometimes called industrial hygienists) are specialists in the assessment of the work environment. They can help managers in the identification, evaluation and control of physical, chemical and other hazards at work. Many people are still unfamiliar with the skills which occupational hygienists have to offer. As a result, they are not always used by managers and supervisors to their best advantage. They can also be used as trouble-shooters when things have gone wrong—a perfectly useful function. However, a better use of their skills can be made at the design stage of any project so as to prevent trouble ever arising.

Threshold limit values

The threshold limit value (TLV) of a substance which can be inhaled is that concentration of the substance to which it is believed that nearly all work people may be exposed repeatedly for an eight-hour day and a forty-hour week without adverse effect. There are several aspects about this definition which require clarification. The TLV represents the best state of present knowledge of what is believed to be safe in terms of known toxicity and irritant effects. The measurement is related to time-weighted average concentrations. It is not, however, a prophesy. Neither is it a dividing line between what is safe and what is unsafe, except where the TLV is qualified with a 'C' or ceiling value (see below). No individual susceptibility or hypersensitivity is assumed, though it is known that a few individuals may experience discomfort or symptoms at or below the TLV. It is also known that pre-existing medical conditions or occupational illnesses may result in symptoms at or below the TLV.

'C' limits of TLVs

In the case of 'C' limits, the TLV is a *ceiling value*, not to be exceeded. The idea of a time-weighted average does not apply. 'C' limits are applied mostly to substances which are fast acting.

TLVs with 'skin' notation

Certain TLVs have the word 'skin' after the name of the substance. This is because *skin absorption*, as well as inhalation, can contribute to overall exposure. Such substances should therefore be kept off the skin in order to prevent general poisoning. The TLV assumes that no skin absorption takes place.

TLVs of mixtures of substances

It is always wise to assume in the absence of evidence to the contrary, that substances may be additive in their toxic effects and that TLVs of mixtures should take into account the worst possibilities. Details of the methods for assessing the hazards of mixtures are given in the reference booklet on TLVs of Airborne Contaminants and Physical Agents published annually by the American Conference of Governmental Industrial Hygienists.

Control

Control of a hazard will be effective if the evaluation has been thorough; control will be less effective if a hasty or superficial attempt only has been made in recognition and evaluation of the hazard(s). Control should follow logically from evaluation.

A recent experience may underline some of the lessons of the method *identification—evaluation—control* as a way of thinking about and tackling occupational hazards of all kinds. An intelligent and able manager was very concerned about cases of dermatitis which were occurring in his factory. In spite of advice from a consultant dermatologist, who treated the cases well, new cases kept occurring and labour relations were becoming soured by the continuance of this apparently intractable problem. The factory employed a part-time doctor and two nurses, none of whom had any training in occupational health, and whose interest was in treating the cases which arose and in doing routine medicals on the staff. At no time had the workplace been properly looked at, nor had the question been asked as to what exactly was getting on to the skins of workers in that particular area and why. When these recognition-of-hazard questions were asked, the problem was quite easily solved: the offending chemical dust was identified and evaluated, and was controlled by being contained in an enclosed system with suitable local exhaust ventilation at points where product inspection or handling was occasionally required. Identification followed by evaluation and control resulted in no further cases.

Education in the hazards of the job

No discussion of control methods for occupational hazards should begin without stressing the very great importance which must be laid on the education of workpeople in the hazards of the job. The best protection that any person can, in general, have about any hazard is a knowledge of that hazard. He should know, for example, how a substance can or could harm him, what he can do to prevent trouble, what has been done in design and engineering of plant and equipment to make the work safe, and the various rules which have to be followed.

The protective value of knowledge—often taken for granted in adults —was brought rather forcibly to the attention of one of the authors recently when he was in a house with a group of immigrant people who knew nothing of gas cookers or of the hazards of gas. They knew that gas could be lit and that somewhere in the business a match was struck and the gas lighted. Unfortunately, they did not have the exact relationship clear. They were charming and intelligent people—but they nearly blew up the house from lack of knowledge!

How many people working in various jobs have an inadequate knowledge of the substances, processes, hazards and risks to which they are exposed, and are thus reduced to the status of children faced with a hazard, or are like the people who had never met a gas cooker? It was necessary in the gas-cooker example to spend time in formal instruction. Nothing else would have been safe or adequate. But do managers and supervisors always give workpeople the elementary protection which knowledge gives? Sometimes, too, concealment is practised—the true dangers may not be mentioned for fear of creating unrest. If a substance is carcinogenic or can kill, people should be told so and this statement should be followed by how to work safely. Co-operation will be forthcoming if the risks are better understood. Deceit will only rebound in time on to the deceivers.

Controlling occupational hazards

Having identified and evaluated any hazards, the methods used to control the problems can be listed briefly *in order of preference of control method*. Before listing the methods there are two questions which should always be raised: 'why do it at all?' and 'why do it in this way?'. If the answers to these questions are reviewed fully, much of relevance will be discussed and subsequent decisions on hazard control, cost, pollution problems and so on will tend to be of better quality.

Hazard control methods in order of preference

The higher the listing of the method, the more it is to be preferred. Several methods will usually be needed to control any hazard.
  1. Substitution of less toxic material.
  2. Enclosure of the harmful process, with automatic operation if possible.
  3. Isolation of the harmful process from the remainder of the plant, with special protection for workpeople necessarily included in the area.
  4. Local exhaust ventilation.
  5. General ventilation.
  6. Wet methods (to control dust).
  7. The use of personal protective devices, particularly respiratory protection.

8. Decreasing the daily exposure through short work periods.
9. Personal hygiene and the use of protective creams.
10. Housekeeping and maintenance.
11. Warnings and publicity.

Substitution of less toxic material

This is the most effective method of dealing with a toxic hazard, the hazardous substance is removed and another substance which will do the job but which does not produce toxic effects is substituted. The following examples illustrate this quite clearly. Carborundum as a material for grinding wheels does not carry the risk of silicosis that used to be found with sandstone wheels. The effects of this substitution on Sheffield knife-grinders in the last century was to prevent many early deaths from silicosis and from the tuberculosis which followed. Today sandstone wheels should not be used.

The dust of asbestos when inhaled can produce diseases of the lung and of the pleura—the membrane which lines the outside of the lungs and the inside of the chest. Many of the present uses of asbestos, such as for low temperature insulation, for roofing material, for domestic iron insulation pads, would best be abandoned because non-toxic materials can just as well do these jobs. For example a new office block was found to have an asbestos-containing textile in the main air intake duct from the plenum ventilation. It appeared that the specification was for a flame-resistant material, but no one had thought to forbid the use of asbestos. The cost of replacing the asbestos was not very high but the completion of the building was delayed at a very considerable indirect cost.

Recently, the use of benzene (C_6H_6) has been prohibited as a solvent because of the high risk of blood disorders from inhaled benzene. Many substances for example, toluene and xylenes, are quite suitable as solvents for applications in which benzene was used. Toluene and xylene must however be benzene-free, because benzene is sometimes found in mixtures with these substances in commercial grades of 'toluene' and 'xylenes'. White spirit used as a solvent is almost hazard-free and may often be substituted.

Carbon tetrachloride is another excellent and cheap solvent which should be replaced by others that are less toxic. A nurse bought some dry cleaning fluid (carbon tetrachloride) to dry clean her wig. The label on the bottle stated in the smallest of print that the substance should be used in a well ventilated room. She cleaned her wig in her bedroom with the window open, but then went to bed and, since it was winter, closed her window. She narrowly escaped with her life since the vapour produced severe liver damage which required treatment in an intensive care unit for over a week.

Enclosure of the harmful process

This is the method adopted in most chemical manufacturing where the product or intermediates are toxic. It is also used to deal with processes involving radiation. Enclosure with *automatic operation* further reduces the risk of exposure of workpeople.

Isolation of the harmful process

This is a useful method for dealing with a hazard in a workshop where most of the work is hazard free. The particular process which is dangerous can be segregated and special precautions can be taken in this area. The workpeople who work in this area may also require special protection. In oil refining, for instance, some additives such as tetraethyl lead and tetramethyl lead are highly toxic. These materials are generally stored in a separate building and special precautions are taken in all operations involving them. The areas where these substances are to be found are isolated from the rest of the plant to minimise hazard.

Local exhaust ventilation

This is the standard way of dealing with dusts, vapours and fumes which arise from point sources of release, for example in grinding, cutting, welding and burning. The harmful substances should be extracted at source so that they do not enter the general atmosphere. The direction of the ventilation should be *away* from the breathing zone of any potentially exposed people. The velocity of air which is needed for the purpose must be adequate and the orifice sizes and the extract fan's capacity to shift the air must be specified exactly. This is a job for a properly trained person. If the calculations leading to a proper specification of air velocity and orifice sizes are not properly made the result will be either inefficient extraction with accompanying risk to health, or a waste of resources and money. Large hoods positioned vaguely and at a distance from the sources of vapours and dusts with feeble fans in large ducts can only be regarded as a gesture and not as an effective extractor. Unfortunately, such do-it-yourself efforts are still quite common. Effective local exhaust ventilation is a very good method of dealing with a dust or vapour hazard, but it must be properly arranged to be effective and thus to prevent illness or disease in workpeople.

General ventilation

The snags with this method are that it is only effective against low concentrations of toxic substances. Large volumes of air may have to be moved and perhaps heated. It is, however, a useful method. One cheap way of ensuring good general ventilation is to build plant and equipment in the open air so that free circulation of air is guaranteed. Inside buildings, the method amounts to opening the windows and doors and moving the air about the room.

General ventilation, which is equivalent to diluting an atmosphere contaminant, can only be used effectively and safely if the following criteria are satisfied:

1. The quantity of the contaminant must not be too great.
2. Workpeople must be far enough away from the contaminant for general ventilation to dilute the contaminant to levels below the TLV, or the contaminant must be evolved in concentrations below the TLV.
3. The contaminant must be of low toxicity.
4. The release of the contaminant must be uniform and without sudden peaks.

General ventilation should only be used, therefore, in those cases where the contaminant is of a low order of toxicity, is produced in large quantities, or is produced from other than point-release sources.

Wet methods

These are used for dust suppression. In coal and rock drilling water is used to prevent dust arising from the drill. A fine spray of water can be used to minimise dust when stripping old asbestos lagging. Powders which give rise to dust when dry will give rise to little or no dust if wetted.

The use of personal protective devices

These devices such as dust masks, ear defenders and gloves are often thought of first instead of last as a method of dealing with an occupational hazard. It is foolish to enclose workpeople in any sort of personal protection when it is the process itself which should be enclosed or when the hazard should be eliminated by substitution. Unfortunately, it is still relatively rare to find people who have not been specifically trained in these matters who think along the lines indicated. Most people, even intelligent ones, seem to regard many hazards as inevitable and without cure, and workpeople as the end victim to be coated in various bits of armour! This may be a caricature, but there is a good deal of truth in it. Another principal drawback to this method is that all protective clothing is uncomfortable or awkward and many people dislike wearing it.

Personal hygiene and the use of protective creams

Washing with soap and water can make a great deal of difference to whether a person develops skin trouble or not. For example, the person who washes seldom, who wears clothing saturated or heavily contaminated with work-soiling and who is generally dirty in his habits, thus increasing the risk of soiling, will be at much greater risk than the fastidious person who takes a pride in his appearance. Cutting oils (oil-water emulsions) which are used to cool the cutting areas of machine tools needed in metal cutting, are a source of potential skin hazards which may vary from irritation and oil acne to, in a few instances, skin cancer. In general, the people who suffer from problems arising from cutting oils use poor work methods, allowing the oil to spray around in a needless fashion, and are not scrupulous about the cleanliness of their skin, their clothes, or the workplace.

Barrier creams are often pushed by sales representatives as a panacea for skin troubles in industry. Far from being so, these creams may easily detract attention from the real needs of the situation which are to keep the offending substance(s) and the skins of the workpeople apart. Substitution, enclosure, better work methods, education, and personal hygiene are all more important and effective ways of tackling the problem. Having said this, and having acted on it, there may be some use for barrier creams. The best thing about barrier creams is that they encourage the user to wash them off at the end of the working stint! Incidentally, the provision of barrier cream is now no defence in common law against claims of negligence.

Housekeeping and maintenance

Tidiness or good housekeeping plays an important role in the prevention of hazards at work. No office, factory installation or plant which looks a mess can possibly be as safe as one which is free from spillages, accumulated dust and dirt, and which is generally clean and tidy. The attitudes engendered in the one when contrasted with the other are bound to have effects far beyond the immediate problems of tidiness.

Good maintenance can also have effects on health. For example, if a dust-extractor fan breaks and is not repaired swiftly, a hazard exists. There is a close relationship between good housekeeping and maintenance, and the attitudes and educational awareness of hazards in any work force. This inter-dependence can be a source of strength or weakness in creating and maintaining safe working conditions.

Warnings and publicity

These are effective as reminders and as stimulators of awareness, provided that education has previously been carried out. Alone, they are at best a very poor way of trying to deal with any hazard, and at worst can be an evasion of a real plan to deal with hazards. Notices and posters which give positive instructions such as 'danger—phenol, wear eye protection at all times in this area' will be more effective than vague exhortations such as 'be safety-minded' or 'look how you go'. Warnings and publicity should always be specific and direct, never vague and general.

Further reading

Occupational Diseases: a guide to their recognition. W. M. Gafafer (ed.), US Government Printing Office, Washington (1966).

Encyclopaedia of Occupational Health and Safety, International Labour Office, Geneva (1971).

Threshold Limit Values of Airborne Contaminants and Physical Agents, American Conference of Governmental Industrial Hygienists, Cincinnati. (Updated and published annually.)

Industrial Ventilation, American Conference of Governmental Industrial Hygienists, Lansing, Michigan (1972).

Industrial Hygiene and Toxicology, 2nd edition, Vol. I: *General Principles,* F. A. Patty (ed.), Interscience Publishers, New York (1958).

The Diseases of Occupations, 4th edition, Donald Hunter, English Universities Press, London (1969).

Occupational Health Practice, R. S. F. Schilling (ed.), Butterworth, London (1973).

Medicine in the Mining Industries, J. M. Rogan (ed.), Heinemann, London (1972).

Industrial Health Technology, Bryan Harvey & Robert Murray, Butterworth, London (1958).

Notes on the Diagnosis of Occupational Diseases, Prescribed under the National Insurance (Industrial Injuries) Act 1965, HMSO, London (1970).

Principles of Medical Statistics, 8th edition, Austin Bradford Hill, The Lancet Limited, London (1966).

Health Hazards of the Human Environment, World Health Organisation, Geneva (1972).

CHAPTER 9

Chemical Hazards

Introduction

Chemical hazards are classified as arising from dust, fumes, vapours, mists or gas.

1. Dust

Dust consists of solid particles which usually arise from crushing, grinding, detonation, impact or precipitation and drying. In still air, dusts tend to settle under the influence of gravity. Toxic dusts can produce lung disease in the form of pneumoconiosis, for example from silica-containing rock and from coal. General poisoning may be produced by inhalation of toxic dust. The engineers of a large oil-burning ship who suffered from vanadium poisoning after boiler cleaning have already been mentioned in this connection. The boiler-soot was found to contain about 5 per cent of vanadium in the form of vanadium pentoxide and inhalation of boiler-soot during cleaning resulted in general poisoning by vanadium. Urinary excretion of high levels of vanadium confirmed the diagnosis. Inhalation of asbestos can, in addition to causing a pneumoconiosis (asbestosis) cause a cancer of the pleura (the lining which wraps the lung and the inside of the chest wall). This form of cancer of the pleura is known as a mesothelioma. Cancer has also been suspected following the inhalation of arsenic dusts. Lung cancer in association with pre-existing asbestosis is virtually only found in smokers. Non-smokers however can get asbestos mesothelioma.

2. Fumes

Fumes are finely particulate solids which arise by condensation from a vapour—often after a metal has become molten. The metal fumes are generally the oxide of that metal, and lead and cadmium, in particular, give rise to highly toxic fumes. Zinc and other metals such as copper and magnesium can also give rise to toxic effects, often in the form of a

Chemical Hazards

feverish illness which follows inhalation of the fumes. This illness is known as metal fume fever. It is quite like the feverish symptoms of influenza and may be mistaken for this. Away from the fumes, the condition usually settles in about forty-eight hours.

3. Mists

Mist consists of fine suspended droplets which arise by condensation from a gas, or from breaking up a liquid by atomizing with compressed air, or in an aerosol, or by splashing, turbulence or foaming in a liquid.

The charging of lead-acid batteries causes hydrogen gas to be given off at the lead plate. The minute bubbles rise to the surface and burst giving off a mist of sulphuric acid. Workers exposed to this acid mist have been found to have damaged teeth since the acid gradually erodes the dental enamel. Acid cleaning of metal also releases an acid mist due to nascent hydrogen. This is usually well-recognised, but if traces of arsenic or antimony exist in the metal then a lethal gas is given off. Such poisoning by arsine or stibine has caused many deaths, often when their presence was completely unsuspected.

Chronic acid mist is given off in chromium plating. If proper precautions are not taken to prevent this mist from reaching the breathing zone of workpeople, nasal ulceration and perforation of the nasal septum can quickly result.

4. Vapours

Vapour is the gaseous form of a substance which is normally found as a solid or as a liquid. Examples of vapours are solvent vapours and petroleum vapours. Mercury also gives off vapour at room temperature, particularly if it has spilled and has broken up into fine droplets thus allowing a larger surface area for vaporisation. Mercury poisoning has arisen from the inhalation of mercury in a laboratory where mercury is lying around or has been spilled. This can be prevented by the use of impervious surfaces and water traps.

5. Gases

Gas is a formless fluid which occupies the space of enclosure and which can be changed to a liquid or solid only by the combined effect of increased pressure and decreased temperature. Examples of the many toxic gases include chlorine, hydrogen sulphide, carbon monoxide and ozone. Many of these are so irritating to the eyes, nose and mouth that people rapidly escape unless they are trapped. More insidious however are carbon monoxide which has no smell and hydrogen sulphide which rapidly leads to loss of consciousness.

Chemical hazards arise from chemical manufacturing, from imported products, from intermediates and from finished products. In every case a specialised knowledge of each chemical is required in order to assess the hazard properly. The principles of dealing with any occupational hazard (p. 72) should be used in relation to chemical hazards. A problem with many chemicals is to identify and to recognise what the substance or mixture is. Often, a manager or supervisor is faced with, for example, a new chemical, such as a cleaner or de-greaser, which is designated by a non-informative name, often that of the manufacturer with some number or letter after it, for example 'Smithclean Z21'. Labelling may or may not be informative, and if it is not informative, the problem must surely be to find out more about the product *before* using it because so many cleaners and de-greasers are known to be toxic by inhalation or on the skin. Many more examples could be given of this important principle. If in doubt, assume the worst and find out. This way is safe.

Dust diseases

1. Silicosis

Silicosis is a pneumoconiosis, caused by dust which is produced by the inhalation of finely divided free silicon dioxide (silica). Silica in a non-free or combined state, as a silicate, will not cause silicosis, though some silicates, such as asbestos and talc, may cause other forms of pneumoconiosis.

Free silica arrives in the lung as dust and the particle size most likely to be retained is about 1 micron in diameter (1 micron is one-thousandth part of a millimetre). The upper and lower limits of particles which will induce fibrosis in the lungs is about 10 to 0·1 microns. Particle shape, surface area and the tendency of any dust to aggregate or not will also influence the capacity to produce lung fibrosis.

Silicosis may develop rapidly or slowly. It most often develops rapidly, within a few years, in people who have been sand-blasting, drilling silica-containing rock, or manufacturing or packing abrasive powders where the alkali content may accelerate the silicotic fibrosis. The outlook for those suffering from silicosis of rapid onset is very poor. The more slowly developing form of silicosis is produced after many years of exposure to silica-containing dust. Workpeople in potteries, foundries, mines and quarries, stone-setting, tile-making, clay-producing and glass-manufacturing may be affected. Radiograms (X-ray pictures) of the chest will usually show characteristic changes. Tuberculosis of the lungs is a common complication of silicosis. Silicosis produces in the victim increasing breathlessness and can lead to a severe crippling of the respiratory system.

2. Coal-workers' pneumoconiosis

Coal-miners are exposed to dust which is a mixture of small amounts of silica and coal. The roles of these and other components of the dust have been discussed at length for many years, many different names have been given to the disease. There is, however, a characteristic set of changes which can be found in the lungs of many coal-miners and which can be demonstrated in chest radiograms. The disease usually takes years to develop, and there is evidence that the removal of affected workpeople from further exposure to the dust of coal mines at an early stage of the disease can prevent the symptoms from worsening, and so prevent disabilities which arise in the later stages. Minor degrees of coal-workers' pneumoconiosis do not shorten life: it is only the more severe forms of the disease which affect longevity. Chronic bronchitis is common in miners but is not recognised as an occupational disease.

3. Asbestosis and mesothelioma

Asbestos is the generic name used to describe different mineral silicates. Most of the asbestos used is *chrysotile*: a simple magnesium silicate. The most dangerous of asbestos is *crocidolite* or blue asbestos. It is, however, probable that *all* forms of asbestos are dangerous to health. Workpeople exposed to high concentrations of asbestos fibres can develop a pneumoconiosis. The onset of the asbestosis is usually related to the amount retained: the higher the dust concentration, the sooner the disease develops. There is evidence that the dust carried home by an asbestos worker on clothing or overalls and the dust inhaled by those living down-wind from an asbestos mine, dump or factory, can, in a small number of people, also lead to asbestos-induced disease in the form of mesothelioma, a form of cancer affecting the pleura or the peritoneum. The pleura, as mentioned earlier, lines the outside of the lungs and the inside of the chest wall; the peritoneum lines the outside of all abdominal organs and the inside of the abdominal wall. Many, if not most, of the present-day uses of asbestos can cause needless exposure of workpeople to asbestos fibres. Substitution of non-asbestos containing materials could and should be more widespread. High-temperature insulation is one of the few uses for which substitution is not available, except ceramic fibre at a much higher cost.

4. Other pneumoconioses

Pneumoconioses can also arise from inhalation of diatomaceous earth dusts (kieselguhr) and from talc. Both diseases are of relatively slow onset and usually requires many years of work with the materials before trouble develops. A benign form of pneumoconiosis can occur from the deposition of iron oxide in the lungs, for example in foundry workers, in welders and in haematite miners.

4. Glass fibre and mineral wool

Both of these materials are now widely used as insulating materials. Glass fibre is made from very fine threads of glass, and mineral wool from molten slag blown into a substance resembling cotton wool by jets of steam or air. There is no evidence that either of these materials has any effects on the lung even though small particles can be inhaled. The only problem that is sometimes reported is a temporary mechanical irritation of the skin in the first week or two of exposure. Thereafter the skin becomes hardened and irritation ceases.

5. Byssinosis

This describes a lung condition caused by the inhalation of plant debris at the early stages in the processing of cotton, hemp and flax. Although it is not a true pneumoconiosis, since fibrosis as such does not occur, it is convenient to include it in this section. The cause of the condition is believed to be an allergic reaction to botanical material, and the main effect is a constriction of the bronchial tubes, similar to asthma. Byssinosis is most commonly found in people who work in the blowing and carding rooms of cotton mills and does not occur once the cotton has been spun into yarn. The usual story is that after a few years at work, a cardroom worker notices that his chest feels tight when he starts work on Monday, and gradually this feeling is experienced on all the days of the week. The end result can be indistinguishable from chronic bronchitis, and many cotton workers all over the world have been permanently disabled by this late stage of byssinosis.

The condition can be prevented by rigorous control of the dust or, as recently suggested, by steaming the raw cotton. Unfortunately, most cotton mills now in operation are old, and adequate dust control of the machinery is extremely difficult. The proportion of workers affected depends upon the amount of dust in the atmosphere, but in carding rooms can range from 5–40 per cent.

6. Bagassosis

This is a condition caused by an allergy of the lungs to spores of fungi which grow in the piles of crushed sugar-cane stalks (bagasse) in the West Indies. The condition is very similar to farmer's lung (p. 154) and should more properly be classified as a biological hazard.

FIGURE 4: *Hazardous dusts and their biological effects*

Dust	Biological effect
quartz (crystalline SiO_2)	silicosis
coal	coal miners lung
graphite	graphite lung
kaolin	kaolinosis
kieselguhr	kieselguhr lung
talc	talcosis
asbestos	asbestosis, mesothelioma
cotton, flax, hemp	byssinosis
hay, straw, grain (contains mould and fungus)	farmer's lung

Other dusts which can give rise to biological effects include aluminium, beryllium, manganese peroxide (MnO_2), chromates, lead, vanadium and basic slag. Deposition and accumulation in the lungs without apparent harm may be identified in those who inhale dust of iron oxides (FeO and Fe_2O_3), barium oxide (BaO) and soot (carbon).

Metal diseases

1. Lead

Lead has probably caused more occupational disease than all the other metals put together. Swallowing lead compounds, for example after storing cider in a lead glazed vessel, can give rise to colic, paralysis and mental disorders. Oxides of lead, when inhaled, will give rise to lead poisoning and in the past this was a potent source of trouble. Inhalation of lead compounds is the usual source of poisoning in industry. Lead paint ('red lead' or 'white lead') is a valuable anti-corrosive, but if metal coated with them is cut by oxy-acetylene torches, oxides of lead may be inhaled with resulting lead poisoning. Ship-breakers have suffered often in this way. Children are very susceptible to the effects of lead and wise parents know the risks involved in allowing children to nibble at or suck anything which may have a lead-based paint. All cots and toys *must* be painted with lead-free paint and any lead or lead-painted objects must be kept away from young children.

At work, people who are frequently exposed to lead, for example, men who manufacture lead compounds, battery makers and grinders of lead-solder in car manufacture, should be monitored for evidence of lead absorption by appropriate medical techniques. It is worth mentioning at this point the difference between the toxicology of the inorganic lead compounds, and the organic lead compounds such as tetramethyl and tetraethyl lead. The way that inorganic lead compounds behave in the body is, with few exceptions, quite different from organic lead compounds because of the completely different metabolic paths. As a result, the signs and symptoms of poisoning by each are quite different. Tetramethyl and tetraethyl lead are used as additives to gasoline. The risk from these substances is generally limited to those who manufacture or work in the petroleum industry. Poisoning by organic lead compounds gives rise to severe mental disorders quite unlike the anaemia, colic and paralysis produced by inorganic lead compounds.

2. Mercury

The saying 'mad as a hatter' originates from the mercury poisoning from which hatters suffered. Mercury vapour from metallic mercury or mercury compounds as dust may be inhaled, and the hatters inhaled mercury nitrate which was used in the carrotting of hair into felt. Inhalation of mercury vapour or mercury compounds is by far the most important route of absorption, but skin absorption can also occur. Police detectives who used to use a fingerprint powder made from mercury and chalk have suffered from mercury poisoning. The symptoms of chronic mercury poisoning are tremor—'hatters shakes'—and mental disturbances in which the person becomes easily upset, timid and emotionally volatile, even losing control of his emotions. Other symptoms include swollen and bleeding gums.

Today spillage of metallic mercury is not recognised often enough as a possible source of mercury poisoning. The metal when spilled will often be subdivided into tiny droplets by being trampled on. This increase in surface area allows increased vaporisation and, in confined spaces or poorly ventilated areas, mercury poisoning can result from inhalation of vapour. Mercury in air is easy to measure and cleansing procedures can render the mercury harmless by converting it to the sulphide. Urinary mercury excretion levels can also be used to check absorption.

Organic mercury compounds differ considerably from inorganic or metallic mercury in their toxic effects. They are chiefly used because of their excellent fungicidal properties and are, therefore, incorporated on wheat and other agricultural seeds as a dressing, in the making of paper and paste, and in marine antifouling paints. Absorption can be by ingestion, inhalation and through intact skin. The alkyl compounds (ethyl- or methyl-mercurys) are more toxic than the aryl salts although their effects are similar. Skin irritation causing a severe dermatitis with blistering can follow contact, but the most serious effects are upon the brain. These effects include tremor, weakness and tingling, leading to

Chemical Hazards 85

permanent damage interfering with speech, hearing and muscular movement.

There have been several episodes of poisoning when treated cereal seeds have been eaten, such as a large-scale outbreak in Iraq, and in the past few years effluent pollution of the sea and rivers which affect fish has caused concern in many countries. The most serious outbreak occurred in the region of Minnemata Bay in Japan. A chemical company discharged mercury into the sea, the compound was changed to an organic form of mercury and concentrated in the shellfish which were a major source of food. There were several deaths. Prevention of poisoning involves applying the usual principles and surveillance of workers with periodic estimations of urinary mercury which will show whether undue quantities are being absorbed.

3. Arsenic and arsine

This favourite of the old-fashioned poisoner is also a well-known occupational hazard. The metal is used as an alloy in metallurgy, and its compounds in glass works, in chemical plants and as pesticides. Arsenic attacks enzymes in the body because it has an affinity for sulphur-containing groups. It is slowly excreted by the urine and is also incorporated in the hair and nails. Acute poisoning in industry is rare, except when substantial quantities have been swallowed. Chronic poisoning affects the skin, produces irritation of the nose, eyes and respiratory tract and may damage the peripheral nerves. Cancers of the skin and lung have developed after absorption of arsenic compounds over many years.

Arsine, or arsinuretted hydrogen, is a colourless gas with a smell similar to garlic, which has caused many deaths at work. It is liberated whenever arsenic or material with an arsenic impurity is acted upon by a strong acid with the release of hydrogen. Episodes of poisoning are usually quite unexpected and have occurred when acid cleaning of arsenic-contaminated metals has taken place. Poisoning by arsine is quite different in its effects from inorganic arsenic. It is usually sudden in onset, occurring within an hour or two of exposure to the gas, and the symptoms are headache, cramps and the passing of blood-stained urine followed by collapse and sometimes death. It is caused by the destructive effect of arsine on the circulating red blood cells. Their haemoglobin is released into the blood, damaging both kidneys and liver. Mild cases of poisoning may be recognised by the development of anaemia and slight jaundice.

4. Cadmium

Cadmium and its compounds are widely used and have caused some deaths and many cases of illness in countries all over the world. Its uses include the protective coating of ferrous metals, in alloys with metals such as copper to increase their strength and wear, as a stabiliser for plastics, as pigments and in heavy-duty alkaline batteries. When absorbed into the body, cadmium is but slowly excreted and it has been estimated that it may take as long as twenty-five years for the body to rid itself of half the absorbed dose. Acute poisoning in industry has most frequently arisen from the inhalation of fumes generated by welding or cutting of cadmium-coated metal. The usual effect is the production of 'metal fume fever', an illness which resembles influenza, produced by inhalation of metallic oxide fumes. If larger amounts of fumes are inhaled this may progress to a severe or even fatal chemical pneumonia. Chronic poisoning can also occur if small quantities are absorbed over a long period. Here the first symptoms may be shortness of breath caused by an irreversible lung damage called emphysema, or urine examination may show a special type of protein caused by otherwise symptomless kidney damage. Serious environmental pollution in Japan a few years ago affected many people whose main complaint was of severe pains in the bones. So troublesome was this that the disease was called 'itai-itai'—the Japanese for 'ouch-ouch'.

The prevention of poisoning requires that great care should be taken to avoid absorption of cadmium. People with previous damage should not be exposed to cadmium and periodic checks of urine will allow anyone showing the protein to be removed from further risk. This is usually sufficient to prevent further damage. Some evidence suggests that men exposed to cadmium for many years may have an increased risk of developing cancer of the prostate gland but this is not yet considered to be conclusive.

Gas diseases

1. Carbon monoxide

This is the commonest cause of fatal industrial gassing accidents, due partly to its ubiquity and partly because it has neither taste nor smell. Carbon monoxide (*CO*) is formed whenever any carbon-containing material—wood, coal, oil, paper and so on—is burned, since full combustion to carbon dioxide is rarely achieved. The proportion of *CO* evolved, however, depends upon a number of factors such as the temperature and the amount of available oxygen. The concentration of *CO* in the exhaust from petrol engines is about 10 per cent when the engine is idling with a rich fuel/air mixture, and 1 per cent or even less at normal running speeds. *CO* is an important constituent of coal gas (but not of natural gas) and is found in blast furnaces and coke ovens, and is often used in the manufacture of chemicals. Studies have shown that

Chemical Hazards

congested city streets do not have dangerous concentrations although these can occur in ill-ventilated tunnels when traffic is halted and engines are left idling.

Carbon monoxide combines with the haemoglobin in blood and prevents the transport of oxygen. The danger of poisoning is thus related to the concentration of CO breathed in and to the duration of exposure.

Early symptoms of poisoning include a headache followed by nausea, dizziness and later these may progress to dimness of vision, coma and death. The TLV is 50 ppm but it has recently been proposed that this should be lowered to 35 ppm. Treatment consists of *immediate* removal from the gas and the administration of oxygen; artificial respiration may be needed if the victim's breathing has become depressed.

2. Carbon dioxide

Carbon dioxide is a colourless, odourless gas with an acid taste which is heavier than air and thus tends to collect in pits and depressions. It is used in the chemical industry, for fire fighting, aerating soft drinks and in the form of 'dry ice' as a refrigerant. It is also produced in large quantities in the fermentation of beer, wine and whisky and also in the decomposition of vegetation. Although it is a natural excretory product of the body's metabolism, a concentration of 3 per cent in the air produces headache and laboured breathing, and at 10 per cent or more, loss of consciousness. It has a toxic action at this concentration even though there may still be enough oxygen in the air. the treatment is *immediate* removal to fresh air and the administration of oxygen. Rigid safety precautions are necessary to prevent serious accidents when cleaning brewing vats or entering sewers.

3. Hydrogen sulphide

Hydrogen sulphide is an extremely poisonous gas with a characteristic smell of rotten eggs in low concentrations. In high concentrations hydrogen sulphide rapidly paralyses the sense of smell. It is formed by the decomposition of some animal proteins and can thus occur in tanneries and sewers. It also occurs in the refining of some metals and is found in substantial amounts of several types of crude oil. The loss of smell makes this gas particularly dangerous. It also has a 'knock down' effect with rapid paralysis of muscles at concentrations of less than 1 per cent.

4. Ozone

A form of oxygen with the formula O_3, this gas is highly poisonous despite the widespread belief that it is 'healthy'. The TLV is only

0·1 ppm and it produces lung irritation at a concentration of 1 ppm. The gas is produced by the action of sparking and high voltage discharges by some photocopying processes using ultra-violet radiation, and by irradiation of air with ultra-violet rays during arc-welding. Inert gas shield welding can produce biologically significant concentrations of ozone within about 1 metre of the arc. Known lung damage has resulted.

5. Oxides of nitrogen

With the exception of nitrous oxide which is used extensively as an anaesthetic the other oxides of nitrogen are all poisonous. These oxides are given off from nitric acid and its action on metals and wood, and are found in the exhaust of diesel engines and in silos. The oxides are also formed by oxidation of the air's nitrogen in the course of electric arc welding or in hot flames, such as oxy-acetylene or oxy-propane, used for cutting metals. The very hot flames used for glass blowing can also give rise to oxides of nitrogen. NO_2 has a TLV of 5 ppm since the gas can cause delayed and long-lasting irritation of the lungs. A latent period of from two to twenty hours between exposure and the onset of serious chest symptoms has frequently added to the difficulty of recognising that the condition was due to this gas. The TLV for NO is 25 ppm.

6. Ammonia

A highly irritating gas with a characteristic smell, which dissolves readily in water and thus irritates all damp body surfaces such as eyes, nose, throat and lungs. Used extensively in chemical processes and as a refrigerant gas in large cooling plants, the main hazard is in the escape of large quantities which overwhelm people before they can run away. Anyone doing maintenance work on ammonia pipes should be wearing suitable eye protection and should have a respirator close at hand in case of any leak.

7. Chlorine

A yellow-coloured highly irritating gas which was the first of the war gases to be used in 1917. It is given off by hypochlorite bleaches in contact with acid solutions and is used in this form or as the gas itself in the chlorination of swimming pools and drinking water. It also has many uses in the chemical industry and can be encountered in many types of factory. With a TLV of only 1 ppm it produces a severe chemical pneumonia which may leave a severely damaged lung after the acute phase has cleared up. The provision of respirators is a necessary safety precaution in the event of an escape of this gas.

Chemical Hazards 89

Vapour diseases

1. Solvents

Many solvents are highly inflammable and this characteristic must not be forgotten even though the main concern here is with health risks that may arise from their use. There are so many available today that one can only make a few generalisations about them. All organic solvents vaporise readily and so are most frequently absorbed by inhalation. Absorption through the intact skin can also, however, take place with many of them. The halogenated solvents, so called because they contain one or more atoms of chlorine, bromine, fluorine or iodine, are not inflammable and thus can be used safely in situations where others can not. Chemically, organic solvents are usually either aliphatic (straight-chain hydrocarbons), aromatic (ring compounds), alcohols, ethers or ketones; some such as white spirit or petroleum ether consist of complicated mixtures of various compounds.

All organic solvents have two health hazards in common: a narcotic effect, like anaesthetics, at high concentration and a damaging effect on skin through the dissolving of natural greases The first can cause fatal accidents and the second can result in skin dryness, irritation or frank dermatitis, and, with a few solvents, a chemical burn. Most aromatic and aliphatic solvents are not particularly toxic, with the notable exception of benzene (C_6H_6). This compound has long been known to cause a severe form of anaemia by damaging the red bone marrow; it is also thought by some to cause a form of leukaemia. For this reason, and since it is very readily absorbed through the intact skin, it should not normally be used as a solvent. Many countries now subscribe to the recommendation accepted by the Council of Europe in 1966 that benzene should not be used as a solvent, and that as an impurity it should not amount to more than 1 per cent of other solvents such as toluene or xylene.

The halogenated aliphatic hydro-carbon solvents vary considerably in toxicity but are frequently used since they are non-flammable. Carbon tetrachloride used to enjoy wide popularity in factories and in the home as an inexpensive and effective solvent, but it can, and does, produce severe liver damage and should not be used as an ordinary solvent (see p. 73). A general purpose non-flammable solvent with negligible toxic properties is methylchloroform (1.1.1. trichloroethane), but as with *all* solvents this has a narcotic effect and can affect the skin.

It would be inappropriate to go into more detail on the wide range of organic solvents. The interested reader should consult specialist literature. It must be emphasised that great care must be taken when selecting a solvent, it should always be used in adequately ventilated conditions and skin contact should be avoided or kept to a minimum. In general those to avoid include: aniline, benzene, carbon bisulphide, carbon tetrachloride, chloroform, methanol, methyl chloride, nitrobenzene and tetrachloroethane.

2. Sniffers

In industry, some volatile substances such as trichloroethylene may lead to addictive sniffing whereby the sniffer gets dreamy and drugged by the substance. Glue sniffing, for the solvent in the glue, has led to deaths following unconsciousness from the anaesthetic effect of such solvents. If anyone is found to be sniffing solvents or other substances they must, for their own safety, be removed at once and for evermore from further risk.

3. Insecticides

Most insecticides in common use have some toxic effect upon man, although pyrethrum, derris, and for practical purposes DDT and gammexane, cause few problems. The biggest headache faced by manufacturers is the remarkable ability of insects to develop resistance, hence the continuing search for new and better insecticides. In use, these substances are either applied in dry powder form or more frequently are sprayed in a solution of kerosene or other solvents such as petroleum naptha (white spirit). In solution, therefore, the general hazards of the solvent apply in addition to those of the insecticide itself. Also, many insecticides can be absorbed directly through intact skin and thus protective clothing is required when spraying is done.

When it became recognised that DDT-resistant strains of insects were increasing, several other chemicals were developed in the group known as the chlorinated hydro-carbons. These include dieldrin, endrin and aldrin which were more persistent in their effects and were most efficient. They had toxic effects upon man but cases of poisoning were infrequent, and took the form of irritation of the brain causing headaches. Heavy exposure resulted in convulsions and coma. Their very persistence, however, posed a serious threat to ecology. These chemicals were stored in the body, particularly in fat and in the brain, and it soon became clear that the bird population was being seriously threatened. The celebrated book *The Silent Spring* by Rachel Carson set out the problem in a most powerful way and the widespread use of these chemicals was controlled by international agreement. Their benefits to man had, however, been considerable since malaria and other insect-borne diseases had been eradicated from many parts of the world largely due to their use.

Chemical Hazards 91

The main group of chemical insecticides in use today are the organo-phosphorus compounds. Developed as an outcome of the study of war gases, these are not so persistent in the body and should be used more frequently than the chlorinated hydrocarbons that they have largely replaced. There is now a wide range of these chemicals including parathion, malathion, TEPP and dichlorvos. They all inhibit the activity of an enzyme in the body called cholinesterase. This is needed to break down the substance, acetylcholine, produced in small quantities by every nerve in the body. Repeated absorption of small doses is cumulative in the short-term and the body's level of cholinesterase is progressively lowered until an acute attack of poisoning can follow. The initial symptom of this can include headaches, faintness, blurred vision and nausea; and in the more severe cases may be followed after a few hours by abdominal cramps, excessive sweating, muscular twitching convulsions and coma. Untreated, such severe cases can die within an hour or so. Fortunately there is an effective antidote which can be given by injection and this, or the use of large doses of atropine, can be life-saving. Protection of the health of people exposed to these organo-phosphorus insecticides consists of careful training, adequate protective clothing and serial estimations of the enzyme cholinesterase in the blood. If the cholinesterase-level drops significantly the workpeople can be removed from further exposure for a week or two until the normal amount is found once again.

4. Herbicides

The killing of unwanted grass or weeds has a wide application in most industrialised societies today. Many substances have been used, some are only mildly toxic whilst others are very hazardous. In every case a careful check should be made on the nature of the chemical involved. Some, such as DNOC (dinitroorthocresol) and pentachlorophenol, are readily absorbed through the skin and can cause a severe generalised illness, and one of the most recent, paraquat, has caused several deaths after accidental drinking, by causing an irreversible destruction of the lung tissue. Other substances can produce dermatitis and these include creosote, trichloracetic acid and maleic hydrazide. The selective or 'hormone' weedkillers which spare grass, have recently been under suspicion of causing genetic damage. This received publicity when they were used as defoliants during the war in Vietnam. Again, it must be emphasized that when using these chemicals care must be taken to follow any safe handling instructions, and when in doubt to consult an expert for advice.

Occupational cancer

Many substances are now known which can produce cancers in man. The resulting tumours may occur at a variety of sites: for example, exposure to coal-tar can lead to skin cancer, and exposure to β-naphthylamine can lead to bladder cancer. Physical agents, too, can produce cancer, for example, by exposure to ionising radiations, by ingestion of radio-active substances and even by long exposure to strong sunlight (ultra-violet radiation).

As time passes, more and more cancers are being shown to be associated with exposure to substances encountered in the environment. Lung cancers associated with cigarette smoking (accounting for two-fifths of all fatal cancers) are today a preventable disease. Yet people still seem willing to expose themselves to an agent which, if the deaths from chronic bronchitis and ischaemic heart disease (coronary thrombosis) are included, is the commonest preventable cause of death and of ill health, and has been recognised as such for nearly two decades Thus, recognition of a hazard, even a cancer hazard, does not necessarily lead to its elimination. In industry, failure to prevent hazards can be followed by legal proceedings and sometimes the recovery of very substantial damages. Recent cases in the courts from the rubber industry and associated with the use of anti-oxidants which led to a bladder cancer, support the view that it is going to be a great deal more costly in future to fail to prevent occupational cancer or the exposure of people to substances which are known to be associated with a higher incidence of cancer.

The general principles of prevention (p. 72) apply to substances capable of producing occupational cancers. A discussion on occupational cancer of the skin is on page 94.

The list below gives some of the individual chemicals or groups of substances which may have been or which may be associated in industry with cancer production.

The list is only of substances which have been proved to induce occupational cancers in man. Many other chemical compounds are known to have induced cancer in experimental animals, but have not, as yet, been proved to induce human cancer. The list of agents given must therefore be incomplete because further search is likely to identify more. A cautious attitude should therefore be maintained, especially towards new substances as they may well be capable of causing cancers in man. A list of agents which *may* induce occupational cancer in man would be very considerably longer. In recent years we have seen the cautious attitude shown towards food additives. Several substances such as cyclamates have been banned because some experiments have shown that they can cause cancer in animals. For various

Chemical Hazards

FIGURE 5: *Examples of carcinogenic agents which have induced occupational cancer*

AGENT	TUMOUR SITE	MODE OF OCCUPATIONAL EXPOSURE
Aromatic amines		
β-naphthylamine	urinary tract (mainly bladder)	production process/use/rubber and chemical industries
α-naphthylamine*		
benzidine		
4-aminobiphenyl	urinary tract (mainly bladder)	manufacture only
auramine		
magenta		
Polycyclic aromatic hydro-carbon mixtures		
coal tar, pitch	skin, scrotum, larynx, lung	production, impregnation and insulation
shale oil, tar	skin, scrotum	petroleum and shale oil refinery
paraffin and petroleum waxes	skin	wax pressing, extracting and refining
soot	skin, scrotum, lung	domestic furnaces, industrial furnaces
anthracene oil	skin, scrotum	purification of anthracene oil and cake
creosote	skin	creosoting lumber
certain mineral oils	skin, scrotum	production work and use
Miscellaneous organic exposures		
isopropanol residue	paranasal sinus, larynx, lung	residue after production of isopropanol
mustard gas	paranasal sinus, larynx, lung	production process, recipients
wood dust, especially hard woods	nasal sinus	furniture making with beech, oak and mahogany
leather dust	nasal sinus	shoe manufacturing
halo-ethers	lung	chemical industry, interaction between formaldehyde, methyl alcohol and hydrochloric acid
vinyl chloride monomer	liver	production of PVC
Inorganic chemicals		
arsenic compounds	skin, lung, liver	viniculture, metal ore, smelting and refining process
asbestos	lung, pleura and peritoneum	asbestos spinning, weaving and insulating
chromates	lung	chromate extraction from ore
haematite	lung	mining
nickel	nasal sinus, lung	refining of metal by the carbonyl process (no cases since 1925)
Radiation		
ultra-violet solar radiation	skin	outdoor workers, e.g., farmers, fishermen, specially white races exposed to strong sunlight
x-ray radiation	skin, bone	cathode-ray and x-ray tube and apparatus manufacture, x-ray diffraction apparatus, roentgenology, radiology, radio-physics, operation of industrially used x-ray, fluoroscopes, diagnostic and therapeutic use, mining of radioactive ore, uranium ore smelters and refineries, mining of other minerals.
alpha, beta and gamma radiation	lung, liver, larynx, thyroid, kidney, subcutaneous and blood forming tissues	

* The carcinogenicity of α-naphthylamine may be due to traces of β-naphthylamine.

reasons the regulations about the use of suspect chemicals at work are much less rigid. Some countries have now forbidden the importation or manufacture of β-naphthylamine, but these are still in the minority. The International Labour Organisation has recently proposed that this and certain other carcinogenic chemicals should be controlled on a world-wide basis.

Occupational skin cancer

The first recognised occupational skin cancer was described in 1775 by Percival Pott, a surgeon at St. Bartholomew's Hospital, London. Pott wrote 'there is a disease peculiar to a certain set of people which has not, to my knowledge, been publicly noticed. I mean chimney sweeper's cancer. It is a disease which makes its first attack on and its appearance in the inferior part of the scrotum; there it produces a superficial, painful, ragged, ill-looking sore, with hard rising edges. The trade call it "soot wart".' Skin cancer due to shale oil was described in 1876 by Joseph Bell, a surgeon in Edinburgh who, incidentally, was thought to have served as a model for the character of Sherlock Holmes. Conan Doyle studied with Bell and admired his deductive reasoning. Shale oil was used extensively in cotton mills as a lubricant and mule spinners were found to develop scrotal cancers.

These cancers were all caused by mixtures of substances. In 1930 Kennway and Heiger first produced a skin cancer in mice by painting their skin with a known pure chemical substance 1,2,5,6-dibenzanthracene. Following this discovery, many pure chemical substances were shown to be capable of causing cancers and a significant step forward was taken in cancer research as a result. Polycyclic (polynuclear) aromatic hydrocarbons, found for example in soot, coal-tar, pitch and mineral oil, are known to cause skin cancer. Ionising radiations and arsenic are also capable of causing skin cancer. People who work outdoors in strong sunlight and who have fair skins have more skin cancers than similar-aged people who have dark skins or who work indoors. Occupational cancers occur more frequently on skin which is normally exposed—face, forearms and hands—or, in the case of men on the scrotum where the baggy, wrinkly skin tends to hold contaminants and thus prolongs exposure. There may also be a particular susceptibility of this rather thin skin to irritant and carcinogenic substance. The importance of good personal hygiene, clean clothes, bathing and washing cannot be overemphasised in relation to the prevention of skin cancers.

Occupational skin diseases

Many skin conditions arising from occupation come under the heading of dermatitis, the inflammation of the skin. The name is apt

Chemical Hazards

to alarm the sufferer because there is a belief that dermatitis is usually or always a chronic and disabling ailment. While this may be so in some cases, many other cases are amenable to simple hygiene—keep the causal agent away from the skin or wash it off quickly and harm will not follow. Much existing occupational dermatitis will clear up simply by removal of skin contamination. This may be done by improved work methods, by changing the sufferer into a different job or by removing him for the time being from work.

Substances which attack the skin may be classified as:

1. Primary irritants

These are substances which, if allowed to come into contact with normal skin in sufficient quantity or for long enough will lead to dermatitis at the area of contact. The mechanism of injury can be de-fatting, precipitation of protein, oxidation or interference with the top keratin layer of the skin. Most occupational skin disease arises from primary irritants.

2. Sensitisers

These are substances which do not cause any demonstrable skin changes on first contact. Repeated or prolonged contact may be necessary before skin problems occur. Contact at one point on the skin may produce sensitisation (allergy) which does not show until further contact with the sensitiser occurs, perhaps at a different place on the skin where an area of skin change is produced. A common example of this is metal sensitivity when a new wrist watch with a nickel strap or back will produce a rash. Previous contact with nickel is required before this can occur.

When normal skin defences have been broken down by irritants or sensitisers and a cracked, oozing and moist skin surface results, secondary bacterial infection can easily occur. Many cases of occupational dermatitis become secondarily infected in this way.

Occupational skin conditions often cannot be diagnosed just by looking at them. A careful history must be taken of the work done, the substances touched or handled, and even those present in the work area, before reaching any conclusions. Patch testing with suspect substances may be necessary. Withdrawal from work may have to be applied as a test in cases of doubt. Occupational diseases should clear up when the person is removed from his work; if it does not, then work is probably not the sole cause and other factors are also operating, or the case is not one of occupational skin disease.

Some of the more important causes of occupational dermatitis are listed below:

1. Irritant dermatitis

(a) Acids: by burning the skin.
(b) Alkalis: by dissolving the skin.
(c) Solvents: by removing the natural grease.
(d) Oils: by blocking the pores.

2. Sensitisation dermatitis

(a) Bichromates.
(b) Epoxy resins and catalysts.
(c) Accelerators and antioxidants.
(d) Many disinfectants or germicides.
(e) Para substituted aromatic amines (e.g., some dyes and local anaesthetics).
(f) Formalin.
(g) Nickel and its compounds.

Detection and measurement of chemical hazards

Although very precise methods to measure concentrations of chemicals in the atmosphere now exist, they usually require sophisticated and expensive equipment such as atomic absorption spectrometers and gas chromatographs. Fortunately it is also possible to obtain reasonable measurements for screening purposes by using small, portable and inexpensive detector tubes. These can be used with the minimum of training and operate on the principle that a colour change occurs in the tube when the specific chemical to which it is sensitive passes through it. The apparatus is simple and robust, and consists of a small hand-operated pump which sucks air through the appropriate detector tube. The approximate concentration is calculated by the length of colouration caused by a specified amount of air, or in another version by the amount of air—number of strokes of the pump—required to produce the colour change. A large range of chemicals can now be measured in this way by using the appropriate detector tubes. The result may then be compared with the recommended safe concentration such as the TLV, and decisions made on the safety of working in the atmosphere concerned. Each tube can, with few exceptions, be used only once, and it is important to note that unused tubes have an expiry date after which their accuracy becomes unreliable.

There are three points about the use of this simple method of measurement which should be emphasised. First, it is not as accurate as those which can be undertaken by a specialist laboratory, and thus

Chemical Hazards 97

there will be occasions when the concentrations may need to be more accurately checked. Second, measurement should be done at the place where a person is likely to breathe the air. Solvent vapours, for example, are heavier than air, and a sample taken at floor level will give higher concentrations than one taken close to the breathing zone of the worker. Third, samples are taken over a few seconds only. The results do not give a time-weighted exposure and should, therefore, be interpreted with caution. However, when used correctly, this simple equipment means that managers, doctors and nurses working in industry can at times make decisions on the safety of work in the presence of known chemical hazards.

Further reading

Encyclopaedia of Occupational Health and Safety, International Labour Office, Geneva (1971).

Environmental and Industrial Health Hazards—a practical guide, R. A. Trevethick, Heinemann, London (1973).

Occupational Health Practice, R. S. F. Schilling (ed.), Butterworth, London (1973).

The Diseases of Occupations, 4th edition, Donald Hunter, English Universities Press, London (1969).

Industrial Ventilation, American Conference of Governmental Industrial Hygienists, Lansing, Michigan (1972).

Industrial Toxicology, L. T. Fairhall, Williams and Wilkins, Baltimore (1969).

Dangerous Properties of Industrial Materials, 3rd edition, N. Irving Sax, Reinhold Book Corporation, New York (1968).

Industrial Hygiene and Toxicology, 2nd edition, Vol. II— *Toxicology,* Frank A. Patty (ed.), Interscience Publishers, New York (1963).

The Merck Index: An Encyclopaedia of Chemicals and Drugs, 8th edition, Paul G. Stecher (ed.), Merck & Co. Inc., USA (1968).

Practical Toxicology of Plastics, Rene Lefaux. Iliffe Books Ltd., London (1968).

Toxicity of Industrial Metals, 2nd edition, Ethel Browning, Butterworth, London (1969).

Medicine in the Mining Industries, J. M. Rogan (ed.), Heinemann, London (1972).

Clinical Aspects of Inhaled Particles, D. C. F. Muir (ed.), Heinemann, London (1972).

IARC Monographs on the Evaluation of Carcinogenic Risk in Chemicals to Man, Vol. I, International Agency for Research on Cancer, World Health Organisation, Lyon (1972).

The Hazards of Work: how to fight them, P. Kinnersley, Pluto Press, London (1974).

CHAPTER 10

Physical Hazards

Physical hazards are often thought to be of less importance than chemical ones. This is not so: they can—and do—cause severe injury or even death. The nature of physical agents is wide and varied and should not be under-rated, and the main ones capable of causing occupational disease or injury are listed as follows: ionising radiations, which include alpha particles, beta rays, gamma rays, x-rays and neutrons, ultra-violet radiation, infra-red radiation, microwaves, laser beams, noise, ultrasound, vibration: repeated motion and repeated impact shock, the hazards of working in high and low temperatures and the hazards of working in high and low air pressures.

Ionising radiations

There are four main types of ionising radiations:

1. *Alpha particles* are high energy particles but because of their size they can only penetrate human tissues to a depth of one-tenth of a millimetre. Irradiation is therefore confined to the immediate vicinity of the source of radiation.

2. *Beta rays* are electrons of various energies and a power of penetration which can travel a few millimetres into human tissues before they are absorbed.

3. *Gamma rays and x-rays* are electromagnetic radiations of varying energies and often of high penetration, capable of irradiating all parts of the body fairly uniformly.

4. *Neutrons* are uncharged particles with a wide range of energy and power of penetration.

Radiation can affect people according to whether it comes from outside the body or from inside. In industry external radiation may occur, for example, from gamma rays or from x-rays produced by sealed sources or from x-ray machines. Internal radiation can result from breathing-in or swallowing radio-active materials. The amount of tissue damage which occurs in people exposed to radiation depends basically

Physical Hazards

on the energy absorbed per unit mass of tissue, that is, on the 'dose' of radiation. Damage also depends on the particular cells, tissues or organs which are exposed to the radiation. Some tissues are much more susceptible to damage by radiation than others. For example, blood forming tissue in the bone marrow and the reproductive cells in the testes and ovaries are very sensitive to radiation. Muscle and bone itself are insensitive. Growing and dividing cells are more radio-sensitive than those which are relatively inactive. The biological effect of radiation of a given amount also depends on the time taken to give that dose of radiation. Low dose rates tend to produce less biological damage than high dose rates.

Radiation absorbed by tissues is measured in *rads*—1 rad corresponding to 100 ergs of energy per gramme of tissue. This amount of energy is about the same as that absorbed when an x-ray dose of 1 roentgen (r) is received by 1 gramme of soft tissue. The biological effect of radiation on tissue depends both on the energy absorbed and on the type of radiation. In order to compare doses of radiation which come from different types of sources but which produce equivalent amounts of damage, the *rem* is used (*r*oentgen *e*quivalent for *m*an).

Industrial uses of ionising radiations

Alpha particles are used in industry to monitor the thickness of paper, polythene films and the like. X-rays and gamma rays are used for metal inspection, and to check welds for faults. X-rays and gamma rays can also be used in absorption devices to detect the presence of certain substances or as level indicators. Radionuclides (radio-active isotopes) have many industrial uses both as a source of radiation, for example, cobalt 60 is used as a source of gamma rays, and as tracer chemicals. Massive gamma-ray sources are used to sterilise medical equipment and food as an alternative to high temperature and pressure which may damage the material or affect taste.

Permissible doses of radiation

Any dose of radiation, however slight, can, in theory, cause some harm and it is impossible to guarantee absolutely no ill-effects. However, to take a practical and not a theoretical viewpoint, certain standards can be laid down as permissible doses for workpeople who are occupationally exposed to radiation.

The International Commission on Radiological Protection currently recommend the following as maximum permissible doses:

1. 3 rems during any period of 13 consecutive weeks to the testes or ovaries, the blood-forming organs and the lenses of the eyes at any age over 18 years (equivalent to 12 rems per annum).

2. 235 rems of 'whole body' radiation for persons occupationally exposed from the age of 18 to the age of 65 (equivalent to 5 rems per annum).

3. 60 rems to the gonads (reproductive organs) by the age of 30 (equivalent to 2 rems per annum).

These recommendations allow for additional radiation from cosmic rays and for a certain amount of radiation from medical procedures. No person should be exposed to more radiation than allowed in any one of the three categories above, they must not be considered independent.

Measuring and recording individual exposure to radiation

It follows from what has been said that levels of radiation must be measured and recorded for each individual potentially exposed to radiation, so that his experience can be compared with the recommended permissible doses. The normal methods of recording doses of radiation are by means of film badges and individual dosimeters. Film badges can, within certain limits, distinguish the kind of radiation to which the badge has been exposed by comparing the film with standards. In this way the dose of irradiation can be estimated. To be useful, the film badge or dosimeter must be worn at all times usually on the jacket where most radiation is likely to occur. Occasionally this may be on a hand. Mistakes in dose rates have occurred when people have taken off their jackets, to which was pinned the film badge, left this in a work area and retired. The film then records a high dose of radiation—but not the dose received by the person. Such mistakes should be avoided by well-trained people. It is also important that records of the radiation dosage of any individual should always go with him. For example, if he changes employer, the records must go with him, otherwise his cumulative exposure to radiation will not be known.

The principles of protection against radiation

Exposure to radiation must be minimised. This can be accomplished by shielding and by keeping distant. The techniques of shielding are a matter for the expert but the principles are clear: the size and direction of the useful beam should be restricted to that which is the minimum required for the work. Scatter radiation which can bounce off hard surfaces should also be minimised in the design of the equipment and shielding. When the radiation beam has done its useful job, the path of the radiation beyond this point must be checked to the point where the level has dropped sufficiently to cause no harm. Cases have occurred of x-ray apparatus being set up in a room with a beam directed at a

Physical Hazards

wall on the other side of which people have been working thus exposing them to the radiation. As in any hazard, the workpeople doing the job must understand the nature of the hazard. Adequate records must also be kept of doses of radiation received by workpeople over many years. Medical examinations are also needed to make sure that people are fit to start this kind of work and that they remain fit to continue in it. In the course of these examinations, special attention should be paid to examination of the blood, including cell counts and blood films.

Many other aspects of protection against radiation such as visual indicators of 'on-off' for x-ray apparatus, storage of isotopes or other sources of radiation, delineation of radiation beams, disposal of radioactive wastes and marking of radiation area boundaries should also be mentioned. However, enough has been said to indicate that protection against radiation is a highly technical subject which needs assistance from people who are familiar with both the physics and the hazards of radiation. Protection of workpeople may also require expert medical assessment. Like most potentially dangerous problems, radiation can be handled successfully and safely if all the rules are followed all the time.

Ultra-violet radiation

Ultra-violet radiation can effect the skin, causing sunburn. A higher incidence of skin cancers is found in the exposed areas of skin of people who are often in sunlight, especially if they have fair skins unprotected by pigmentation. Certain other changes which resemble those of age in the skin are also produced by ultra-violet radiation and these are often seen in farmers, sailors and other people who work outside.

Conjunctivitis can also be caused by ultra-violet light and is commonly called 'welder's flash' or 'arc eye'. Coal-tar, creosote and pitch are photosensitising agents which promote the action of ultra-violet radiation on the skin. There must be many who have learnt this themselves after creosoting a fence in summer.

Infra-red radiation

Infra-red radiation is felt on the skin as heat and so normal defence mechanisms act to protect the skin from burning. The eyes, however, can suffer damaging amounts of radiation without much feeling of heat. The effect of repeated exposure to infra-red radiation on the eyes is to cause a cataract (an opacity of the lens), in this case an opacity of the posterior part of the lens. This is known by a variety of names such as 'glass blower's cataract' or 'chain maker's cataract'. Foundry workers and men in constant daily contact with hot metal, glass, or people working in kilns may be at risk. Dark glasses will protect the eyes from this form of cataract by cutting down the infra-red radiation reaching the eye.

Microwaves and radiofrequency waves

Like infra-red radiation, microwave radiation produces heat, but it is a great deal more penetrating, and this characteristic has led to the development of microwave ovens now coming into widespread use in industry. At frequencies below 30 megaHertz, which represent radio and radar wavelengths, the heating effect becomes negligible and they do not represent a hazard to the health of operators.

Despite some alarming articles that have appeared in the press on the hazards of microwave radiation these are very similar to those long since known for infra-red radiation. Apart from a very noticeable sensation of heat felt under the skin rather than on its surface, the organs mainly at risk are the cornea and lens of the eye since they do not have a blood supply which can dissipate the heat. A safe exposure limit of microwave radiation has been agreed internationally at 10 milliwatts per square centimetre, and interlocks are necessary on oven doors to ensure that radiation ceases when the door is opened. Leakages of radiation should not exceed 5 milliwatts per square centimetre and a notice should warn operators not to place their faces near an oven that is switched on. In other respects the risks are similar to infra-red rays and the sensation of heat produced is enough to warn the unwary. Microwave and radiowave generators, however, all require very high voltages and thus there is also the well-known hazard of electrocution.

Laser beams

A laser beam is a beam of light energy (photons) of one length (monochromatic) and uniform in phase; the waves travel together in step and in rhythm (the oscillations are synchronised). The name laser derives from *l*ight *a*mplification by *s*timulated *e*mission of *r*adiation. Laser beams have a number of applications in cutting and welding, in photography and in a variety of other fields such as medicine, for instance to spot-weld a detached retina, and in communications.

The energy of a laser beam can cause biological damage and the organ most readily affected is the eye. Protection is based on avoidance of the laser beam and its reflection. Inadvertent operation of the laser is particularly hazardous. Eye protection is advisable to keep down the energy dose to the retina of the eye. The rooms in which lasers are operated should have black non-reflective surfaces to cut down reflection.

Physical Hazards

Noise

Noise exposure can be defined as the total acoustic stimulus applied to the ear over a period of time. Noise emission is an index of the total noise reaching the ear over a specified period of time. This is usually expressed as the noise emission level in decibels (dB). In measuring noise a sound-level meter is used which can measure the sound-pressure level of any noise emission. The human ear responds in a non-uniform way to different sound-pressure levels, that is, it responds not to the time intensity or real loudness of a sound of given frequency (pitch), but to the perceived intensity or apparent loudness. A weighting curve, called curve A, has been constructed which takes into account these differences. Sound-pressure levels are therefore expressed in dB(A), that is in decibels conforming to weighting curve A, because this reflects the perception of that sound emission by the normal human ear.

Industrial noise

The effects of noise in an industrial context can be described under two main headings: the effects of noise on workpeople and the effects of noise on the community.

The effects of noise on workpeople

Exposure to noise at work or elsewhere may result in loss of hearing by noise-induced hearing loss. Impact noise such as is made by hammering or riveting can produce noise-induced hearing loss in a shorter time than most other industrial noise. Occasionally, gunfire or other explosive noise can result in immediate deafness. However, occupational deafness is generally a slow insidious deterioration of hearing over a period of years. Other effects of noise will include the psychological ones of annoyance, of irritation at a particular noise, loss of concentration produced by intermittent noise, and difficulty in working in environments where the sound level, though not producing hearing damage, is sufficiently loud to make normal communication difficult.

Continuous and discontinuous noise

Measuring and evaluating exposure to continuous noise is relatively easy. However, discontinuous noise—especially if it is variable— is much more difficult to quantify. Methods have been devised to convert measurements of component parts of noise for given times into an equivalent continuous sound level expressed in db(A). A nomogram for doing this appears in the code of practice for reducing the exposure of employed persons to noise (see *Further reading*). Another method of assessing exposure to discontinuous noise is to use a personal noise dosimeter.

Personal noise dosimeters

Instruments are available which can measure and summate the noise to which an individual has been exposed. This information may be specially valuable in the case of workpeople who move around in areas where noise is of differing intensities and where it is difficult to know exactly how long they spend in each area. If a personal dosimeter is worn over a representative period of time the risk can be assessed.

The effect of noise on the community

Those who live near noise-emitting sources, whether these be airports, factories, heavy traffic, railway lines or power stations, react towards the source of the noise in a variety of ways. For example, the drivers of large lorries may be less motivated to complain of the noise from lorries than a resident of a quiet village through which the noisy lorry passes. The area in which the noise is being made, the time of day of the noise and whether the noise is continuous, constant in intensity, or discontinuous will all tend to influence the nuisance and complaint ratings. Some noises may be so loud as to cause health hazards from deafness; others may be of nuisance value only.

Measuring noise

Noise emission levels are levels measured by a sound-level meter and are usually recorded in dB(A). An overall level of 90 dB(A) is presumed to be the maximum acceptable level for people to work in for eight hours a day in a reasonably steady sound without producing noise induced hearing loss. This level is not a desirable level but the maximum level without wearing personal ear protection. Each increment of 3 dB(A) above 90 is presumed to halve the time to which the person can be exposed to the noise without long-term danger to hearing. So, a sound of 93 dB(A) should not be heard for more than 4 hours, a sound of 96 dB(A) for more than 2 hours, a sound of 99 dB(A) for more than 1 hour on any day, and so on.

Noise may also be measured in octave bands. The resultant plot shows the 'sound spectrum' and indicates the characteristics of the noise, whether mainly high-pitched, low-pitched, or of variable pitch, or whether one particular component of the sound in a narrow band dominates the overall picture. This information is useful in engineering control and can also be plotted in such a way as to show presumed safe time limits of exposure to the noise. The effects of attentuation by ear muffs can also be shown on the chart. In this way, a record can be used to show workpeople the risks of damage when their ears are unprotected and the beneficial effect of hearing protection.

Having recognised that noise can produce hearing loss the problem can be evaluated, and control can be planned. As usual in these sort of problems it is both cheaper and more effective to begin at the design stage in any project to ensure that plant and equipment, offices, computer rooms and restaurants have sound levels which are appropriate to their function (p. 133). The sound level should also be such that it will not lead to hearing-damage or psychological ill-effects in workpeople.

Design consideration in regard to noise

It is specially important that noise should be considered in the design stages of plant, equipment and buildings because it may be that this is the only point at which really useful and significant steps can be taken to prevent high noise levels. Unfortunately, it is still relatively unusual for machinery to be ordered from manufacturers with a stipulation that the noise emitted from the machinery should not exceed certain levels. Similarly, many offices and other buildings are designed without heed for the desirable noise levels inside. For example, it is a great deal easier and less costly to fit double glazing and to increase the mass of walls to reduce noise at the design stage, than it is to do it later.

Dealing with existing noise

Much can be done to reduce noise emission at its source even if the building or machinery is already in use. Sound insulation depends on mass. High-density materials will often be used, therefore, in the construction of sound-insulation baffles and walls. Resonance and reflection, which may add to the general level of noise, can be reduced by the judicial use of non-resonant mountings for equipment, or by adding baffle boards or walls in the critical places. Double glazing, the use of acoustic tiles or carpeting may also be necessary. It must be emphasised, however, that modifications such as these require expert advice based upon scientific measurement since much money can be wasted by the over-enthusiastic. Acoustic tiles are useless for sound insulation. They are only useful in cutting-down reflected sound. Properly applied, these methods are much more effective than placing reliance on the wearing of ear plugs or muffs; they are less at the mercy of human behaviour even if the initial expenditure is greater.

Work limitation

This may be a method of controlling a noise problem. People can be exposed for limited periods below that which is presumed to be capable of causing damage (p. 104, damage risk criteria) and audiometric changes (p. 107).

Hearing-conservation programme

A hearing-conservation programme is a formal way of dealing with the prevention of noise-induced hearing loss. It can also be used to include problems of community noise because much of the work of measuring noise levels for the prevention of hearing loss within a factory can be used to build up knowledge of noise sources which may affect the community around the factory.

The steps to be taken in any hearing conservation programme are to measure noise, evaluate the information and control noise exposure by: (i) *Design* of plant, equipment and buildings; (ii) *Engineering methods* such as enclosure of noise sources and attention to noise generation sources; (iii) *Personal protection*.

Personal protective devices

Undoubtedly the best method of noise attenuation for personal protection is a set of good ear muffs, but where lesser degrees of attenuation are required, glass-wool used as earplugs or individually-made earmoulds may be perfectly satisfactory. The sound level at the workplace will to some extent determine which method to use. For noise of over 100 dB(A) ear-muffs will be required. Levels of 90–100 dB(A) can usually be sufficiently attenuated by ear moulds or well-fitting ear plugs. Between 80–90 dB(A) glass-wool will be effective. It should be mentioned here that cotton-wool is *useless* for protection against noise and thus should never be used. If personal protective devices of any kind are to be worn they must be acceptable to those who use them. Any form of protection which is supplied but not worn is useless and is also a waste of money. Acceptance by the wearer will significantly increase the periods when the devices are worn. Even a choice of colour among approved ear-muffs may make a difference to wearability.

Audiometry

Hearing can only be tested properly by using an audiometer. This instrument uses pure tone sounds of differing frequencies and the subject is asked to indicate when he hears the sound. Audiometry must be done in a sound proof room or booth. In this way an audiogram is pro-

Physical Hazards

duced. An audiogram shows the sound pressure level (volume) measured in decibels (dB) at which the subject hears a sound of given frequency (pitch) measured in Hertz (Hz). If the volume of the sound has to be raised in order that the subject should hear it, the added volume is shown below the zero line at the top which represents normal hearing. Characteristic patterns are shown by the normal falling in hearing acuity which comes with age (presbycusis), by noise-induced hearing loss and by a variety of other conditions which are of medical interest. Noise-induced hearing loss shows a characteristic dip in the curve at the 4000 Hz frequency which may also be seen at 3000–6000 Hz and in greater hearing loss may also spread more widely (see figure 10, p. 147).

A classification of noise in the community is available which recognises the variables and which attempts to lay down standards for various different situations. The Wilson Committee which reported on noise in 1963 suggested that the noise levels inside living rooms and bedrooms, which should not be exceeded for more than 10 per cent of the time are:

	Day	Night
Country area	40 dB(A)	30 dB(A)
Surburban areas away from main traffic routes	45 dB(A)	35 dB(A)
Busy urban areas	50 dB(A)	35 dB(A)

Where speech communication is important, the upper limit of noise inside buildings should be 55 dB(A).

Ultrasound (ultrasonics)

Sounds of over 20,000 Hz are not heard by the adult human ear and are called ultrasound. Local heat is caused in tissues exposed closely to sources of ultrasound. Liquids are caused to form cavities by ultrasound and when the cavities break up, shock waves result which can tear holes in metals. Blood may also do this and the resulting shock waves can cause cellular and tissue disintegration. The risks to people however, are not great unless ultrasound is propagated through a liquid in which the individual or a part of him is immersed.

Vibration

The vibrations which are of interest are in the range 30–400 Hz. Under 3 Hz the whole body vibrates and any effects which may occur will be those of motion sickness. Over 400 Hz the vibrations behave as low sound. Different parts of the body resonate at different frequencies, for example, the skull resonates at 30–90 Hz and the consequent vibration of the eyes will upset vision.

In industry, the common effects of vibration are seen in association with the use of vibrating tools such as power saws, pneumatic drills and hammers, and other power tools. People who use these tools may find that they suffer from white fingers and a numb 'dead' hand. The onset of white fingers and numbness is usually aggravated by cold. This disorder is indistinguishable from the condition called Raynaud's phenomenon which is caused by a localised spasm of the arteries to the fingers. Raynaud's phenomenon affects women more often than men, and although it can be very uncomfortable, like vibration-induced white fingers, it causes no lasting ill-effects. Some people appear more sensitive to the effects of using power tools than others and studies of chainsaw operators in forestry work, for instance, have shown that up to one-third of the people at risk can be affected after several years' work. It is said that the condition does not cause disablement but sufferers find that they can no longer hold ice-cold glasses of beer in the affected hand. Degenerative changes in bone and white fingers are more often seen in people who are using heavy low-speed tools rather than light high-speed tools.

The hazards of working in extreme temperatures

The human body has a very sensitive temperature control mechanism which is a great deal more efficient than most thermostat-controlled central heating systems. Although skin temperature can vary by many degrees, the deep body, or 'core', temperature remains remarkably constant at 37°C (98·6°F). Heat is generated throughout life by the chemical processes of metabolism, and this internal heat is greatly increased by muscular exertion. Heat and cold represent opposite extremes of body temperature, and as an occupational hazard excessive heat is a great deal more common. All sources of energy used today produce heat, even though this may merely represent an inefficient waste of power. Many machines, and computer or other electronic equipment in particular, produce a great deal of unwanted heat, and special cooling plants may be required for the sake of the equipment more than for reasons of human comfort.

Heat loss from the body

The control of body temperature is maintained by regulating the loss of heat continuously being produced. This loss takes place in three main ways: by radiation, by convection and by evaporation of sweat. In a comfortable environment most of the heat is lost by radiation and by convection, but when the core temperature starts to rise the skin is moistened with sweat. Evaporation, by the process known as latent heat loss, then aids cooling and this is *by far* the most efficient method

Physical Hazards

of losing heat. In industry, when conditions exist with substantial amounts of radiant and convected heat, the air becomes warm and sweating is necessary for the body to lose the heat gained from without as well as from within. The air close to the body soon becomes saturated from the evaporation of sweat and thus air movement is necessary to maintain heat loss. In some factory processes heat and high humidity are both necessary, as in textile spinning or in some chemical or biological manufacturing processes. Limitation of the time exposed to such conditions may then be the only safe way to preserve a normal body temperature. Human sweat contains a substantial amount of salt and unless this is replaced a body deficit of salt can develop, which in itself may aggravate the effects of heat.

Acclimatisation occurs in people who move to hot climates and, to a lesser extent, in those exposed at work on a regular basis. This means that after about two weeks they can lose heat more quickly and efficiently with less salt loss. Recent research has shown that acclimatisation can occur by regular daily exposures in a climatic chamber and this procedure is of value in military occupational medicine in that it enables soldiers to be more rapidly effective when they are flown from a temperate to a hot climate. Clothing is also important as most man-made fibres such as nylon do not absorb sweat. Heat is therefore **not** lost as efficiently as when natural cotton is worn.

Heat illness

Failure of the body to adjust to heat stress produces illness in two ways:

1. A depletion of salt due to excessive loss in sweat, or when insufficient salt is taken by mouth to replace it. This is particularly likely when strenuous work is done in a hot environment by one who has not become acclimatised. Undue tiredness, nausea and muscle cramps may ensue, leading to a condition known as 'heat exhaustion'.

2. A rise in body temperature due to failure of the normal cooling mechanisms. This is uncommon except in extremely high environmental temperatures such as the deserts of the Middle East. Sweating ceases, core temperature rises and when it reaches about 40°C (104°F) convulsions may occur. At even higher body temperatures coma and death follow.

Prevention of heat illness

Work people exposed to hot environmental conditions should be physically fit and, if possible, become acclimatised to it. An adequate intake of fluid and salt must be maintained. A man may require up to 15 litres of fluid and 30 grammes of salt per day. Salt can be added to water or fruit squashes at a concentration of one gram per litre without affecting palatability. Clothing should be loose and permeable to allow air to circulate close to the skin, but may need to be dense enough to provide insulation from radiant heat. When exposure to high radiant heat loads is inevitable, rest periods in a cool room will be necessary. Specially ventilated and cooled protective clothing may also be needed. Physical exertion should be kept to the minimum to reduce internal heat production. If workpeople are exposed to high environmental temperatures, it will be the manager's responsibility to obtain professional advice on safe exposure times which can be calculated when measurements have been taken. This is yet another example of the systematic approach: identify, evaluate and control.

Work in cold environments

In outdoor occupations such as fishing or farming work must often be undertaken in very cold climatic conditions. This is also necessary indoors, in such places as cold stores for food, or in test rooms where equipment may be run in arctic conditions. The general effects of cold are to produce excessive heat loss and thus a fall in core temperature. The natural body reaction is to increase heat production by muscular activity as in shivering, but this is very tiring and is only effective for a limited time. Heat loss is further aggravated by moisture on the skin and thus wet cold is more dangerous than dry cold. Cold can also produce local damage to fingers, toes and nose in the form of 'frostbite', caused by freezing of the tissues and with an end result very similar to a burn.

Preventing cold stress

The selection of work people can be important: those who are thin and droopy tend to be more sensitive since they lack the insulation provided by subcutaneous fat. Moderately fat and muscular people with an extrovert disposition seem, from industrial experience, to be the best to select. Protective clothing is essential to insulate the body and a thin layer of warm air trapped in the interstices of a string vest provides good insulation. Outer clothing should, of course, be windproof and waterproof but sufficiently permeable to allow sweat to evaporate. Extremities also require protection from frost bite and a loss of dexterity is the price that must be paid. Where exposure to cold is continuous, adequate rest periods must be arranged in warm rooms where hot drinks and food can be taken to provide both heat and metabolic fuel for protection against chilling.

The hazards of working in high air pressures

High air pressures are experienced by those who work in compressed air, for example in tunneling, in diving, and in caissons. The effect of very high environmental air pressure is to increase the amount of air dissolved in the blood and body fluids, although this in itself causes little trouble until the person returns to usual pressure levels. The drop in pressure releases the gases from solution just as when the cap is removed from a soda water bottle. The bubbles consist almost entirely of nitrogen since oxygen is used up by the body. These bubbles then cause 'the bends' or *decompression sickness,* a condition characterised by dull pain in the muscles and joints and bones. Another effect of these nitrogen bubbles is to block the small arteries supplying certain bones giving rise to a condition known as *aseptic bone necrosis,* in which local areas of bone die. This rarely causes symptoms at the time but characteristic changes can be seen in bone radiographs. Ear, sinus and tooth pain can also occur during decompression when air expands in the hollow cavity. All these conditions can arise if decompression takes place too quickly.

Any person who has to work in compressed air must be medically examined before work starts and must be surveyed for fitness at intervals. No-one with a cold or with any respiratory infection should be compressed or decompressed as this may give rise to serious ear damage with bleeding into the ear drum. It is also necessary to carry an identifying tag to show that the person is working in compressed air and giving the address of the nearest decompression chamber in case of decompression sickness. Any person who suffers from any effect of decompression must be immediately re-compressed. Then, very slow decompression is commenced watching carefully that the symptoms do not recur.

Tables of recommended decompression times must always be used in compressed air work. These give the times and rates of decompression following various durations and degrees of compression, both for divers and for workers in compressed air.

The hazards of working in low air pressures

Aircraft pilots or others exposed to high altitudes without pressurised cabins may suffer from the effects of gas bubbles in their tissues, but this is not common. The commonest effect of altitude is oxygen lack in the rarified air. Lack of oxygen can lead to blacking-out or to behavioural changes. Oxygen masks must be worn in aircraft at altitudes above 10,000 feet unless the cabin is pressurised. Without cabin pressurisation, the atmospheric pressure at around 40,000 feet becomes so low that even breathing pure oxygen is not enough to keep the brain and body supplied. This fact was insufficiently appreciated by the early aviators and was the cause of several fatal accidents.

Further reading

Encyclopaedia of Occupational Health and Safety, International Labour Office, Geneva (1971).

Environmental and Industrial Health Hazards: a practical guide, R. A. Trevethick, Heinemann, London (1973).

Occupational Health Practice, R. S. F. Schilling (ed.), Butterworth, London (1973).

The Diseases of Occupations, 4th edition, Donald Hunter, English Universities Press, London (1969).

Health Hazards of the Human Environment, World Health Organisation, Geneva (1972).

Industrial Hygiene and Toxicology, 2nd edition, vol. 1—*General Principles,* F. A. Patty (ed.), Interscience Publishers, New York (1958).

The Hazards to Man of Nuclear and Allied Radiations, HMSO, London (1963).

Noise (The Wilson Report), HMSO, London (1963).

Hearing and Noise in Industry, W. Burns & D. W. Robinson, HMSO, London (1970).

Occupational Hearing Loss, D. W. Robinson (ed.), Academic Press, London and New York (1971).

Code of Practice for Reducing Exposure of Employed Persons to Noise, HMSO, London (1972).

Guide to the Safety Aspects of Human Vibration Experiments, Draft for Developments, No. 23, British Standards Institution, London (1973).

CHAPTER 11

Safety at Work—
Preventing Injuries and Damage

Incidents, injuries and damage

An 'accident' may be defined as 'an event without apparent cause'. In colloquial terms, however, the use of the word 'accident' with its overtones of the supernatural, the unpreventable and the attitude of 'I am blameless as far as causes of the events leading to injury are concerned', is a means of excusing, condoning and failing to face up to what is known about the causes of incidents, injuries and property damage and to what can be done to prevent them from happening. Thus what most people mean when they talk of an 'accident', is either an incident of an unfortunate nature or an injury or both.

In reality, however, experience shows that:

1. Most injuries are the result of *unsafe acts,* often by the person who is injured, less often by others.

2. Most so-called *accidents are non-accidental.* A chain of events precedes the injury or damage, the links of which are made up of faulty and unsafe acts, and of unsafe conditions.

3. An *identification* of the problem, together with a *measurement* and *evaluation* of the data produced, can lead to successful preventive action and to *control* of the problem.

This chapter must, therefore, start with our strongly-felt conviction that in any work situation, the number of personal injuries and the amount of property damage which occur can be reduced by the use of existing knowledge and methods. There is seldom any mystery about *how* and *why* most injuries and damage occur: the causes are usually obvious and, in nearly every case, can be found by investigation. When these causes are clearly identified and are presented in numerical terms, their relative importance can be assessed. Preventive effort can thus be directed towards those areas which will give the most important pay-off at that point in time. As the effort in one area succeeds, other areas will assume a greater importance. The relative priorities should be noted in order of importance in a continuous way so that overall direction is given to the progress of the injury and damage prevention programme.

A *safety programme* can and should be formulated and directed by setting objectives. Events should not dictate progress towards goals: the manager's task is to work towards influencing events in a given direction.

Personal injuries and property damage are forms of *inefficiency*. There should be no need to labour the personal suffering which accrues to an injured person and to his family; it should be self-evident. Unfortunately, in our society today, this is not always so. It is salutary to think for a moment of what would happen if one were suddenly the victim of some easily preventable incident which resulted in a personal injury, the consequences of which included an operation, followed by a five-month's stay in hospital, a further three months' off work and an end result involving permanent curtailment of leisure activity. This is not a fanciful image; such injuries—and worse—occur daily at home, on the roads and at work. But, *almost all* are preventable, and the vast majority are easily preventable. Most are therefore unnecessary.

A few injuries are fatal, some result in loss of time from work but most are less severe. The same could be said of the economic loss from property damage. The damage is catastrophic in a few cases, very costly in some, but usually the cost of any particular incident is not large. Cumulatively, however, the cost is very large.

Generally stated, the manager's task is to influence the course of events in certain directions. To be party to any unnecessary form of suffering or inefficiency is to condone it. The purpose of this chapter is to show that a manager's actions, or lack of action, can materially influence, upwards or downwards, the number of injuries and the amount of property damage which occur in the areas which he controls. In some companies the safety performance of any individual manager is regarded as an easily identified index of his general capacity to manage, and generally this seems to be true. Well-managed organisations usually have a good safety performance whereas badly-run organisations have a poor safety record.

Do you have a problem?

The first stage in any problem solving is to identify clearly what the problem is about. While this seems an obvious statement, the underlying premise that there must be awareness of the problem being there at all is not. A simple example may serve to illustrate. Having lived in a room every day does one notice the deterioration in the paintwork and the wear in the fabrics in the same way as another person who enters the room for the first time? Untidiness may, for example, be regarded as normal. Poor lighting, bad work practices, unsafe acts, non-compliance with safety regulations and so on may be accepted without

Safety at Work—Preventing Injuries and Damage 115

question. It may, therefore, be advisable to try to view the situation afresh or to get an outside opinion if there is some doubt about the existence of a problem.

Another way of answering the question 'do you have a problem?' may be to compare the safety performance of the organisation or group with others. The form of comparison may be international, national or local in score, with a similar industry. As long as measurable indices can be produced, then comparisons can be made.

Measuring safety performance

Measurements which could be used in the assessment of safety performance are:

1. The number and kinds of injuries (fatal, time losing, requiring medical treatment) per day, month or year.

2. The number and kind of incidents which give rise to property damage (catastrophic, serious, minor) per day, month or year.

3. The numbers and kinds of unplanned emergencies which occur.

4. The cost of property damage over a month or year.

Injuries are often measured as a *frequency rate*: the number of injuries in a given period divided by the total number of man-hours worked in that period and the result expressed as injuries per 100,000 or 1,000,000 man hours. Working hours lost due to injuries are also calculated in a similar fashion producing a *severity rate*. However, leaving aside all details of which measurements are made and what the individual measurements mean, the act of measuring and recording is, in its own right, a vital key to knowledge in any problem, whether the problem is production, accounting, research or safety, and knowledge which is in the form of numbers is more precise, useful and comparable than opinions or impressions. An aim of any good safety programme must be to have an adequate flow of information from which problems can be recognised, progress charted and future action planned, and which can answer the question, 'have we a problem and what size and shape is the problem?'.

Basic preventive theory

To be effective, prevention must seek to remove causes. This means that *all* contributory causal factors must be sought and identified clearly, and that each must be eliminated. Injuries and damage are the end results of the chain of causal events. For example, if a car is in an uncontrollable skid, the end results may vary from death of the occupants and writing-off the car, to a small bump on the car or to nothing at all. But in each case the only way to prevent the consequences is to prevent the skid.

In any chain of causes, there eventually come a point when prevention can no longer be applied and where chance will determine whether the consequences are fatal injury, catastrophic damage, serious injury plus damage, minor injury and damage or no injury plus damage, damage and no injury, or finally no injury and no damage. It also follows that if a skid results in no injury or damage it should not be regarded as an unimportant event: the potential consequences of such a skid are death of the car occupants and total loss of the vehicle. In any one particular instance, the consequences, *by chance,* may not result in death or destruction. If, however, we fail to use such an incident as a warning and do not seek to prevent the skidding in future, *a valuable preventive opportunity has been ignored or missed,* which could save lives, prevent serious injury, avert catastrophic damage and save the cost of car repairs.

Similarly, in an industrial context, should a heavy weight fall unexpectedly, various combinations of injury and damage occur. The incident in each case is the same, the weight falls suddenly and unexpectedly; the consequences in each case are different. Once the weight begins to fall, it is *chance* which determines the consequences. Therefore preventive effort, to be effective, must be applied *before* the weight falls. Once the weight is on its way down, prevention is no longer possible. It is necessary, therefore, to examine the chain of casual events which precede the fall of the weight if we are to prevent the injuries and damage.

The fall of the weight is called the *critical event*: critical in the sense that prevention can no longer be applied after the event has taken place. In the examples above, the critical events are the car skidding and the weight falling.

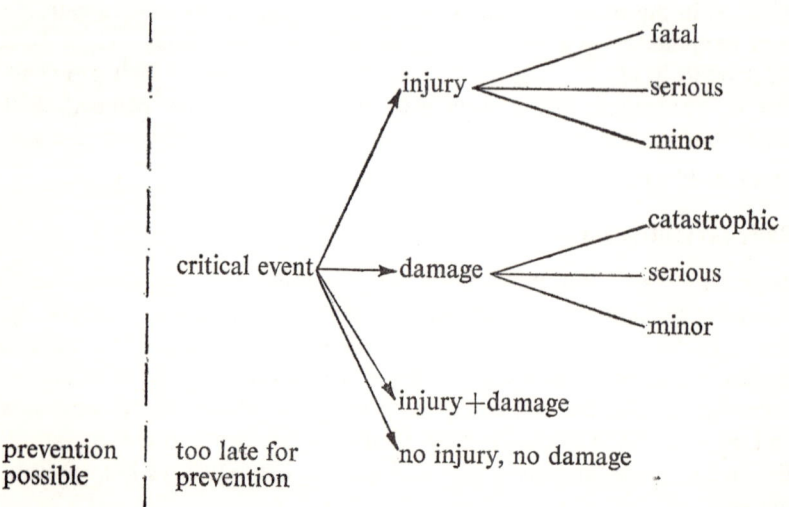

```
                              injury  ─── fatal
                                      ─── serious
                                      ─── minor

  critical event ─── damage  ─── catastrophic
                             ─── serious
                             ─── minor

                    injury+damage
                    no injury, no damage

prevention   | too late for
possible     | prevention
```

In order to prevent the consequences, or possible consequences, of the critical event, it will be necessary to examine the incidents and events which led up to the critical event and to cut the chain of causation at one or several points so that the critical event will not take place.

Why did the critical event take place?

In the case of the weight falling, there may have been failure of the suspending wire, or mechanical failure on the crane (environmental causes, *unsafe conditions*); or there may have been inattention or improper operating procedures on the part of the crane operator (personal causes, *unsafe acts*); or perhaps a combination of these conditions, for example, improper operating procedure, inattention leading to mechanical failure of the crane and severance of the suspending wire.

In the case of the car skidding, the skid may have been due to worn tyres or defective brakes or to oil on the road surface (environmental factors, *unsafe conditions*); or the skid may have been due to taking a corner too fast or to bad application of the brakes or to bad driving by a person whose blood-alcohol level was high (personal factors, *unsafe acts*). A combination of all these factors may produce a chain of events which could be described as follows: an inexperienced youth, whose blood-alcohol level is high, drives a powerful car with which he is not familiar, at night on wet roads, at too high a speed. On approaching a corner, which he does not see in time due to the defective headlamps on the car, he takes the wrong line, brakes violently on an oily surface and goes into a skid. The front tyres of the car are subsequently found to be worn and inflated to an unsuitable pressure.

It is probable that *all* the factors named in the above chain of events had some bearing on the production of the critical event, in other words, causation is multiple. Some causes may be more important than others but all probably contributed to the final skid. Each contributory factor which is unrecognised and which is not corrected is a preventive opportunity missed.

Causes are multiple

There is no such thing as *the* cause of any incident. Modern preventive theory has demonstrated in many fields the validity of multi-factorial causation and to speak of *the* cause of any incident or injury is both to be inexact and, more important, to overlook some factors in causation which may be turned to preventive gain. Failure to recognise a contributory causal factor will mean that this factor cannot be the subject of preventive effort, and thus may continue to exert its effect in leading to further incidents, injuries and damage. Recognition of the importance of every causal factor will lead to efforts to eliminate each of these causes, preferably with some ideas of the magnitude and relative importance of each casual factor, so that priorities can be established in attempting to deal effectively with each factor.

The sequence of causal events leading to injury, damage, both or neither, whether at home, on the roads, or at work, could be shown diagramatically as follows:

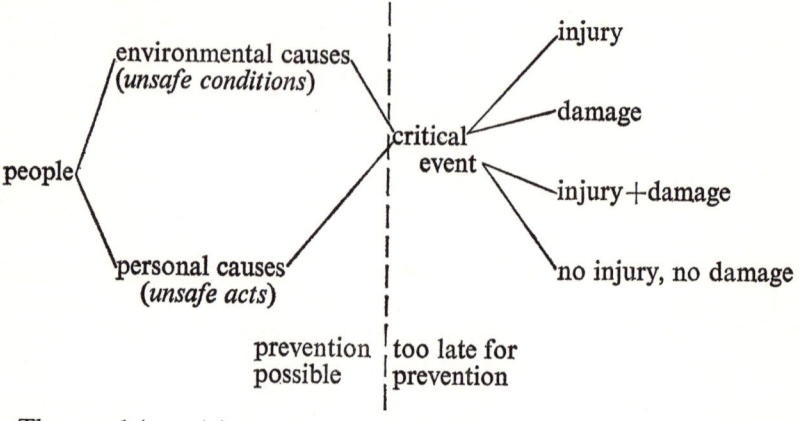

```
                                              injury
         environmental causes
         (unsafe conditions)
                                              damage
                              critical
people                         event
                                              injury+damage

         personal causes
         (unsafe acts)
                                              no injury, no damage


             prevention   too late for
             possible     prevention
```

The word 'people' appears at the beginning of the chain because it is people who create safe or unsafe environments and it is people who carry out safe or unsafe acts.

Unsafe acts and unsafe conditions

When the causes of incidents, injuries and damage are analysed into major groups of unsafe acts and unsafe conditions, it will be found that most causes are related to unsafe acts (usually 65–85 per cent) and that unsafe conditions contribute to a much lesser extent (about 15–35 per cent).

Each of these large groups should, of course, be broken down into smaller and more precise sub-groups so that specific remedies can be sought.

For example, if the unsafe act is 'failure to wear appropriate protective equipment' the remedies may be to supply acceptable equipment or to educate the workforce in the use of protective equipment or to discipline a man who has been told repeatedly to wear protective equipment and has failed in this instance to carry out his job instructions—and so on. It is not enough to categorise the unsafe act without understanding *why* the unsafe act was carried out and *what remedy* is appropriate. Unless such details are sought, we get such statements as 'piece of metal flew off a chisel hitting man in the eye' as a statement of *cause*. The statement is perfectly correct as a *description* of an *incident* but is quite nonsensical and useless for preventive purposes. It does not

Safety at Work—Preventing Injuries and Damage 119

say *why* the metal flew off or *what remedy* is appropriate, for example, whether eye protection should normally be worn on such a job, whether the job was being correctly carried out, or whether the head of the particular chisel had 'mushroomed' and should never have been used at all. All of these factors must be considered, and the unsafe acts and unsafe conditions sorted out and identified for remedial action.

Human behaviour will be found to be the major causal factor in human injuries, for even unsafe equipment can be due to human failings.

Mode, place, time and prevalence

In order to complete our ideas about causes it is necessary to record in every instance some information under each of the above general headings.

1. *Mode* will explain the 'hows' of the incident and should describe in terms of adjectives or adverbs what happened. The descriptions should use words like 'suddenly' or 'without warning' or 'slowly'.

2. *Place* should always be described in detail.

3. *Time* must be recorded as it may have important bearings on all other casual factors.

4. *Prevalence* is the notion that certain events are linked. For example, if it is found that nearly all the cases of sprained ankles occur in men at around noon and nearly always adjacent to a patch of round ground beside a refectory it can be inferred that some particular activity involving men prior to lunch time is related to sprained ankles—and this may either be rushing over rough ground or playing football. If the prevalence is not noted, the events may not seem to be connected in any way and the opportunity to prevent will be missed.

Applying prevention

It will therefore be apparent that an exact categorisation of causes, a quantification of each contributory cause and an analysis of the relevance of these in the production of incidents, injuries and damage is the foundation of successful direction of preventive activities. Prevention should thus follow logically from an understanding of causation and must, in point of time, be applied *before* the critical event takes place in order to be effective.

Analyses of incidents, injuries and property damage in terms of environmental and personal factors all show that personal factors account for a greater proportion of the causation than environmental factors. Our prevention efforts, on the other hand, tend to be directed more to the environment than to the people. Without minimising any present efforts in the environmental field, we believe that it is necessary to try to get across a better understanding of the theory of causation, for without this it is not possible to show the importance of the person and his behaviour as causal factors. Omission from our thinking of any part of causation, which is multiple, means that an opportunity to prevent has been missed.

We should therefore study the methods by which human behaviour can be influenced—by education, leadership, training, discipline, propaganda, example, exhortation and so on—and weigh up the effectiveness of each method in any situation. We must also cease to neglect or ignore the contributions towards safety which can be made by psychologists, educators, epidemiologists, bio-statisticians, and other experts in human capabilities and limitations. For it is in the field of the person and his behaviour that the major part of the causation of most injuries, and damage lies.

Influencing human behaviour

Human behaviour is the major contributor to the causes of incidents, injuries and damage, whether arising directly from unsafe acts or arising indirectly from the creation or toleration of unsafe conditions. It therefore follows that we must seek to influence it. Much is known about human behaviour, attitude formation, motivation and indeed about a host of factors which influence the way people cause incidents, injuries and damage. Unfortunately much of this knowledge is neglected or ignored. The reasons are probably many and varied, but two are given below:

1. It is always easier to suggest that things in the environment were 'the cause'. Things will not answer back (or argue) if it is decided that they are mainly responsible for certain events, or for the consequential injuries or damage. People are understandably less willing to accept that their actions contributed to such undesirable consequences and will certainly defend themselves if they feel that they are being blamed.

2. The training of many people, including managers, does not include the behavioural sciences; more commonly the background is technical—either commercial or scientific. Much of their work is concerned with mechanical, engineering, accounting, production and technological problems. It is therefore not surprising that they see injuries and damage mainly in environmental terms, without analysing the underlying motivating factors. There is, of course, no doubt that they contribute causally in a major way to about 20 per cent of all injuries and damage. In the remaining 80 per cent of cases, however, human behaviour is the major cause—and therefore this must be examined and corrective action undertaken.

Formulating a safety programme

The objective of any good safety programme is to ensure that all members of the organisation work safely as a normal part of carrying out their jobs. Good safety practices, safe working methods and safe, tidy environmental conditions must be an integral part of the ordinary work methods. If this objective is to be attained, a safety plan should be set out so that everyone knows the plan, how the plan is to be carried out, and knows their own role in carrying out the plan.

A good safety programme will have two main parts:

1. *Management principles.* These principles must be explicit and given sufficient publicity. Everyone in the organisation must know and understand that top management is committed and expects the organisation's aims to be directed to safe working. Safety begins at the top: without this, everything else will tend to fail.

2. *Detailed plans.* This part should highlight numerical targets, specific problems causing most injuries and damage, the measurements of progress against targets, and plans for improving safety performance, including training and education.

Some basic management principles for safety

Safety is a line responsibility
Concern for safe working and the belief that incidents which lead to injury and damage are preventable phenomena must begin at the top. If those at the highest levels of management do not show enough concern for the life, limb and health of the people whom they supervise, then the performance of those under them will show a similar lack of concern. It will follow that all managers and supervisors must be held directly responsible for personal injuries which occur in those they supervise, and for damage to plant and equipment which is in their charge.

A corollary of this view is that any safety personnel in the organisation must perform a staff function. That is, they must be safety *advisers* to the line management. They will have important functions regarding the collection and distribution of information, suggesting appropriate safety goals, much leg work on a day-to-day basis to help the safety programme along, and the provision of an advisory service to line management; *but,* they must remain as advisers. If the line manager or supervisor thinks that somebody else is doing the job, he will assume that he is not responsible and the consequences will be unfortunate.

Safety should *not* be the job of the specialist. It is organisationally permissible to have an adviser in a staff function on any problem including safety, but it still remains the work of the line management to turn out work on time, in the right quantity, of the right quality, without a high percentage of rejects, spoiled work and so on, *and* without injuries or damage. This responsibility belongs with the line management and should not be removed from it.

Attitudes are important

How people act depends on what they believe. Attitudes reflect beliefs and behaviour is motivated. People do not do things for no reason at all. Because the causes of most injuries are due to human actions, whether as 'unsafe acts' or due to people creating 'unsafe conditions', it follows that a great deal of effort must be directed towards influencing human motivation and behaviour. This is not always thought of as an important part of any safety programme. Indeed, a caricature of a 'traditional' safety programme might be that the manager considers putting guards on a machine and thinks about sticking a few posters in the work areas.

It is our belief that the great majority of managers and supervisors concentrate too much on the environmental-mechanical aspects of safety problems and not nearly enough on the behavioural aspects. We do not question the sincerity of their efforts and accept that some effort must be directed towards the environment to prevent injuries and damage. But we do question the lack of emphasis on human behaviour and attitudes, leading to unsafe acts as causes of incidents, injuries and damage. Often 90 per cent of the available effort is directed towards the environment and towards correcting unsafe conditions—which we know to be important—but this only covers about 20 per cent of the problem. The contributions of mechanical engineers and chemists, to name but two specialists, have been relatively well exploited in the control of environmental hazards. However, the potential contribution of specialists in human behaviour to safety has been largely neglected. We hope that this will not continue for much longer.

What are existing attitudes?

The attitudes of managers and supervisors are, like other people, governed by what they believe. It may therefore be necessary in order to understand or to change existing attitudes to discuss generally within management what level of safety performance is the right one for the organisation and for the individual manager or supervisor at any particular point in time. How many deaths represent an acceptable norm? How many deaths, time-losing injuries and all injuries per year are within acceptable limits? How much property damage? How much inefficient use of human and material sources by these wastages is to be condoned or accepted? What amount of human suffering and loss of life or limb, which is known to be largely controllable and prevent-

Safety at Work—Preventing Injuries and Damage 123

able, is to be accepted as inevitable, and what is to be a realistic goal for improvement next year?

Conscious awareness of the management attitudes to these questions can provide a good starting point for the formulation of objectives in the prevention of incidents, injuries and damage. As with other problems, objective setting is a means of studying a problem and is a technique which can aid progress towards realistic goals.

Objective setting must be by mutual consent: imposed or unrealistic targets can only bring the system into disrepute. On the other hand, careful addition of incremental improvements in small groups, each of which has its own targets, into an organisation overall total will gain commitment and acceptance. Progress against targets can then be used for a genuine aim of management by objectives.

Why investigate?

The careful investigation of *all* incidents which give rise to injury or to damage should be a cornerstone of any preventive programme. All dangerous occurrences which come to light should also be investigated, but many will not be reported, human nature being what it is! However, the aim should remain of casting the net as widely as possible, for it is only in this way that more preventive opportunities can be created.

The question is often asked 'why bother to investigate trivial injuries or incidents in which neither injury nor damage resulted?' The answer is that *by chance* the circumstances of the incident led to these terminal events; but in other circumstances the consequences could well have been very serious. For example, we have seen a very trivial cut on a man's little finger. This injury was due to the man gripping a handrail when he slipped while he was 120 feet off the ground. He was left hanging one-handed on this rail. Here was a potentially fatal injury which was averted only by a strong grip! The severity of the injury bore little relationship to the severity of the event. If the injury and the incident had not been investigated the state of the man's footwear and the inadequacy of the guardrails during a construction period would not have come to light. Immediate preventive action ensured that a similar incident could not occur. Failure to report or investigate any unexpected or untoward incident results in missed opportunities for prevention.

Who should investigate?

In every case the supervisor of the injured person or the damaged equipment should investigate. This follows the principle of line responsibility. If the incident is more serious, for example, if an injury causes time off work, a local rule might be made that the supervisor must seek the assistance of the safety adviser.

How should investigation be done?

Several points should always be made clear in any investigation:

1. The purpose of investigation is to seek causes and thus to apply preventive action.

2. The apportionment of blame is not a function of this kind of investigation.

3. *All* causal factors must be sought.

4. The investigation must result in suitable practical recommendations which can be followed up and put into effect in order to prevent similar incident/injury/damage.

5. The quality of investigation which is carried out, the nature of the recommendations and their translation into practical terms must be reviewed with the supervisor who completes the investigation form by supervisors or managers who are one or two levels higher in the organisation.

6. The information on the completed form is used to compile statistical information about the numbers and causes of incidents/injuries/damage which occur, and that this information is made generally available to guide future preventive action.

A specimen of a suitable form for investigating and reporting on incidents, injuries and damage is given in Appendix II (p. 161).

Follow-up

There must be adequate follow-up of the recommendations of the investigation so that future incidents are prevented. Similarly, the experience of all incidents/injuries can be presented statistically both to show what has happened over a period and more important to guide future preventive action.

Presenting the results of investigation

The results of investigation must be presented in an interesting and informative way to those who need or can use the information. A mixture of narrative, statistical tables, trend graphs and histograms is usually better than too much of any one. In presenting trend graphs, for example the numbers of all injuries and of time-losing injuries by months, twelve-month moving averages give a much more informative trend line than actual monthly figures.

Each organisation will have means of communication which are used for various purposes: the important thing with information such as we have been discussing is that it *is* communicated and that it is not stored away in a place where only a few people see it. The more pertinent the information and the better it is presented, the more chance has it of being used to change attitudes and to produce action.

Incentives

Some may think it strange that incentives are offered to persuade people to do what is undoubtedly in their own interest. A view often expressed is that incentives are in some way immoral and that the idea itself is wrong. Another way of looking at the problem of using incentives might be to ask 'do they work?'. The answer is 'sometimes' or 'not always', but *used appropriately,* there is no doubt that incentive schemes can *help* a safety programme along. Used in the wrong way or in isolation from other safety effort, the result may be to turn attitudes in the wrong direction. They also tend to lose their effect after a time and may also become expected by staff as a right and even as a 'fringe benefit'.

To avoid any possible misunderstanding, the authors do *not* consider the payment of 'danger money', 'height money' and so on to be either an incentive or a proper cause for payment. No self-respecting manager or supervisor should ever ask any other person to carry out unsafe work and bribe him for doing so. Work should be made safe as far as possible.

In summary, the main ideas in this chapter are:

1. That fostering safe working has to do mostly with *people* and with *attitudes*. Therefore, preventive effort must be strongly orientated towards *influencing human behaviour* in the appropriate directions

2. That influencing the environment by guarding machinery, by using technical expertise to produce safe conditions for working and so on is an important *part* of the total problem. But it is only a part.

3. That legal minima in applying prevention will only produce minimal effect.

4. That, in spite of what people may appear to believe, the environment is *not* the sole injury and damage producer.

5. That stating the activity in which a man was engaged when he was injured is *not* the same as giving the causes of the injury or damage.

6. That people are the most important causal factors of human injuries and of property damage.

If preventive efforts are to succeed, much greater attention must be directed towards the behaviour aspects of the problem and towards producing social, cultural, and individual attitude changes.

By working towards such changes as a deliberate act of policy, management can produce more at less cost, measured both in lessening of human misery, and in increased efficiency, because injuries and damage are forms of inefficiency.

Further reading

Accident Prevention Manual for Industrial Operations, National Safety Council, Chicago (1969).

A Review of the Industrial Accident Research Literature, A. R. Hale & M. Hale of the National Institute of Industrial Psychology, HMSO, London (1972).

2000 Accidents: a shop floor study of their causes, P. I. Powell, M. Hale, J. Marton & M. Simon, Report No. 21, National Institute of Industrial Psychology (1971).

A Cost-effectiveness Approach to Industrial Safety, T. Craig Sinclair, Committee of Safety and Health and Work Research Paper, HMSO, London (1972).

Safety and Health at Work, Report of the Committee 1970–72, Chairman Lord Robens, Cmnd. 5034, HMSO, London (1972).

CHAPTER 12

The Working Environment

What kind of working environment is wanted?

Many working environments, whether offices, workshops or whatever, are pleasant, clean, and tidy. They contain the necessary physical equipment for the work task, such as telephones, hand tools or elaborate machinery. They also have the sort of things which people require in order to feel human and comfortable. For example in an office, the office furniture is pleasant, the seats are comfortable, the decoration is pleasant, and the atmosphere of the office indicates that the management care about the people who work in it. Similar examples could be given about many other working environments and about maximising the things which make it pleasant for people to work in that environment.

Attention in this way to the work environment will cost something—not necessarily a lot. But it will probably bring a return in terms of better work because people feel that they are being cared for and will enjoy working in such an environment. By contrast, workplaces can be needlessly unpleasant for lack of care, thought and cleanliness. Such an atmosphere promotes untidiness, with resultant inefficient work and needless injuries, and a 'don't care' attitude on the part of people who are treated in this way.

We believe that there is a positive gain in better work, in better human relations, and in happiness when people work in pleasant surroundings. If the job has of necessity to be done under unpleasant conditions of one kind or another, there is no need to have locker rooms, eating places, lavatories and so on similarly unpleasant. If anything, these should receive special attention to make them as pleasant as possible. The work environment can be made to express quite a lot about how a management group feels about the human beings who work in that area. The feedback from the workpeople can be positive or negative, both in terms of work and of human relations. It is therefore worth thinking about the type of feedback which is desirable and about how this could be helped by attention to the human qualities and the aesthetics of the work environment.

Ergonomics and design procedures

Ergonomics has been defined as the study of the relationship between man and his working environment. It is perhaps best thought of in terms of the man-machine interface. Good ergonomics maximises the potential contribution of both the human and the machine to the system. The history of machinery shows that work systems were often designed with little thought for the human operator. He was, for example, apparently expected to operate controls situated above his head while looking downwards at work in progress—and the control often went from left to right when the work did the reverse. Such anomalies were scarcely helpful to the human operator trying to do a good job. Thought was seldom given to those parts of the operation which could best be done by the machinery and those which could only be done by people. Man is after all a very inefficient source of power but excels at taking decisions and dealing with mechanical breakdown. It came to be realised that a study of normal human qualities of visual acuity, length of reach, ability to respond to auditory and visual signals, temperature comfort zones and so on could guide design engineers to produce better controls so that the man–machine interface did not cause unnecessary difficulty and working environments were controlled to improve rather than hamper working efficiency. The battle is, however, far from won. The instruments and switches on the fascia of most motor cars and the range of movement and support given by the driving seat often show elementary faults in design. Instruments can take time to read, time better spent on watching the road ahead. Wipers, lights and other equipment switches may be indistinguishable and a long drive in most driving seats may be a sure recipe for backache.

A simple example can be used to illustrate the kinds of approach which could be described as traditional and those which are good in terms of human design engineering. Let us suppose that six instrument dials have to be surveyed and monitored by a human operator. Two of the dials refer to one area of machinery and the other four refer to another area. In the first layout, all the instrument dials are seen giving a normal reading. They are grouped in two tidy rows of three. Each one has to be read individually to check that it is normal and then, if it is not normal, the instrument has to be related to the part of the machinery to which it refers and which is not functioning properly. To complicate matters still further, the overhead lighting may also produce a reflection from the glass covering some or all of the dials.

The Working Environment

FIGURE 6: *A comparison between two methods of instrument groupings*

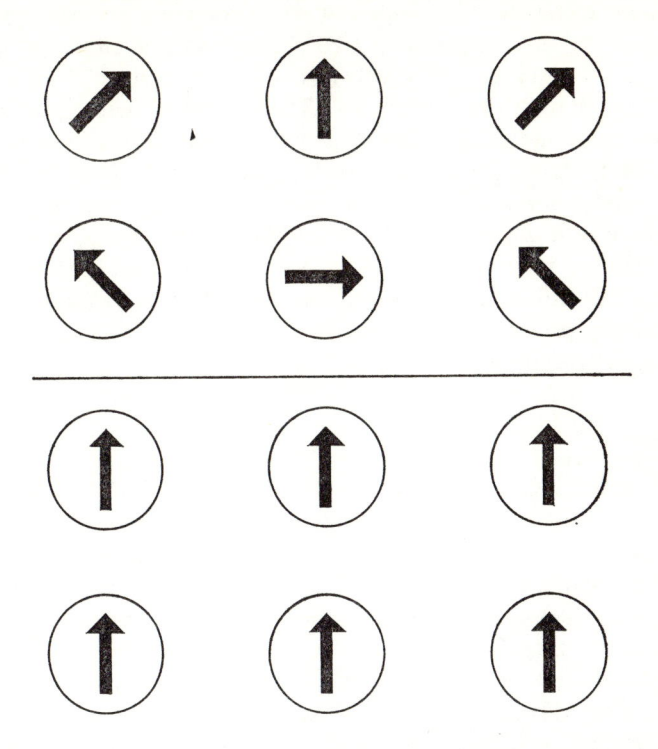

In the second layout, the instruments are grouped according to the part of the machinery to which they relate. All the dials are set to point in one direction when the readings are normal. In this way all the instruments can be read at a glance. The dials are also clearly illuminated without glare and non-reflective glass may be used.

Many examples could be given of the importance of design in the field of safety. Aircraft crashes, collisions at sea and many other disasters have sometimes been shown to have had faulty design as a factor in their causation. These examples are given to underline the importance of trying to achieve good design at the beginning, so that human beings can more easily and without error perform their role in the system.

Optimising the physical working environment

Comfort in any working environment will depend on temperature, air movement and humidity, on the level of illumination and its relationship to visual efficiency, on the noise level, on the comfort of seating, on the ease of performing tasks with machines and other equipment, and on the aesthetics of the environment. There is no doubt that the ease and efficiency with which work is carried out can be readily influenced by any of the above considerations, too much or too little heat, for example, probably will lead to discomfort. The right amount will result in maximum efficiency. The design of desks, seating, work benches, machine tools, control panels, visual information and so on can lead to tasks being performed easily or with difficulty. Some of these problems will be discussed in more detail below. Aesthetic considerations are easily missed out of discussions about environments, with needless ugliness as a result. In terms of cost it is often not a matter of spending money. But it does require extra thought to produce aesthetically pleasant surroundings and it does require that the thinking be done at the design stage.

Lighting

People see better when there is more light. Dark work, that is, work with dark materials or in dark places usually needs more light. The finer the detail of the work the higher will be the level of illumination required. People also see better when the work they are doing is lit so as to be in contrast to its surroundings, usually by being brighter. The background of the work should in general not be highly polished to prevent reflected glare. Care should also be taken to conceal light sources from direct vision so that they are not a source of glare. Large windows and skylights may need blinds to cut out glare. Artificial lighting should be of satisfactory colour and intensity and, if fluorescent tubes are used, flicker must be avoided. Most sources of light generate heat: this should be considered in design, because it could be a problem. Ceilings should reflect light. To do this, the colour should be light so that at least 70 per cent of light striking the ceiling is reflected. Walls should not be too dark and, except near large windows, should reflect at least 40 per cent of light.

Last, lighting cannot be planned in isolation from the aesthetic considerations involved in creating a satisfactory work environment. Visual efficiency can result from good lighting, but even with good lighting, the total environment must be aesthetically pleasing for the best effect. For example, any contrasts in lighting must be of the right kind; colour must be appropriately and sensibly used; and other environmental factors such as heating and noise must be within acceptable limits. Lighting is only one of the many factors which can make or mar a good environment.

Recommended levels of overall illumination have been rising steadily over the years. Much difference of opinion exists about the scientific basis for the determination (or guessing) of these levels. However, in

The Working Environment

practice the proposition that 'people see better when there is more light' with which this section started, is a sound one. A few suggested minimal illumination levels are given below as guides:

	Minimal level of illumination in lux* (lx)
Office work	400
Fine bench work	600
Exacting visual tasks	1,000
Very fine work	2,000
Corridors and stores	200
Places seldom visited but requiring to be lit for safety	100

*1 lux=1 lumen per square metre (1 lumen per square foot=10·76 lux).

The thermal environment—heating/cooling, humidity and ventilation

These are all basic considerations in any working environment. Comfort requires that the moisture content of the air should not normally rise above about 75 per cent or fall below 20 per cent, and that ventilation should permit 2 to 4 air changes per hour if sedentary work is being done. More changes may be required if the work is hot.

The temperature of the work environment will depend on the task to be performed. Some guides are given below:

Sedentary work	19·0–21·0°C (65–70°F)
Light work	15·5–19·0°C (60–65°F)
Heavy work	13·0–15·5°C (55–60°F)

Mental activities begin to slow, errors and accidents increase when the temperature rises above 30·0°C (86°F). Stiffness in extremities begins when the temperature drops below 10·0°C (50°F) and there is good evidence to show that construction workers in the winter work more slowly and have more accidents when they are cold.

One of the biggest problems when deciding just what values are required for the thermal environment in a place of work is the wide range of human preferences. What to some people is a pleasant sensation of freshness induced by air movement is considered an uncomfortable draught by others. Most recent research papers on the subject now take care to describe conditions that suit 80 per cent of the people. This is one of the problems that arise in large offices and, to paraphrase Abraham Lincoln, it seems impossible to satisfy all the people all the time.

FIGURE 7: *Noise rating (NR) curves*

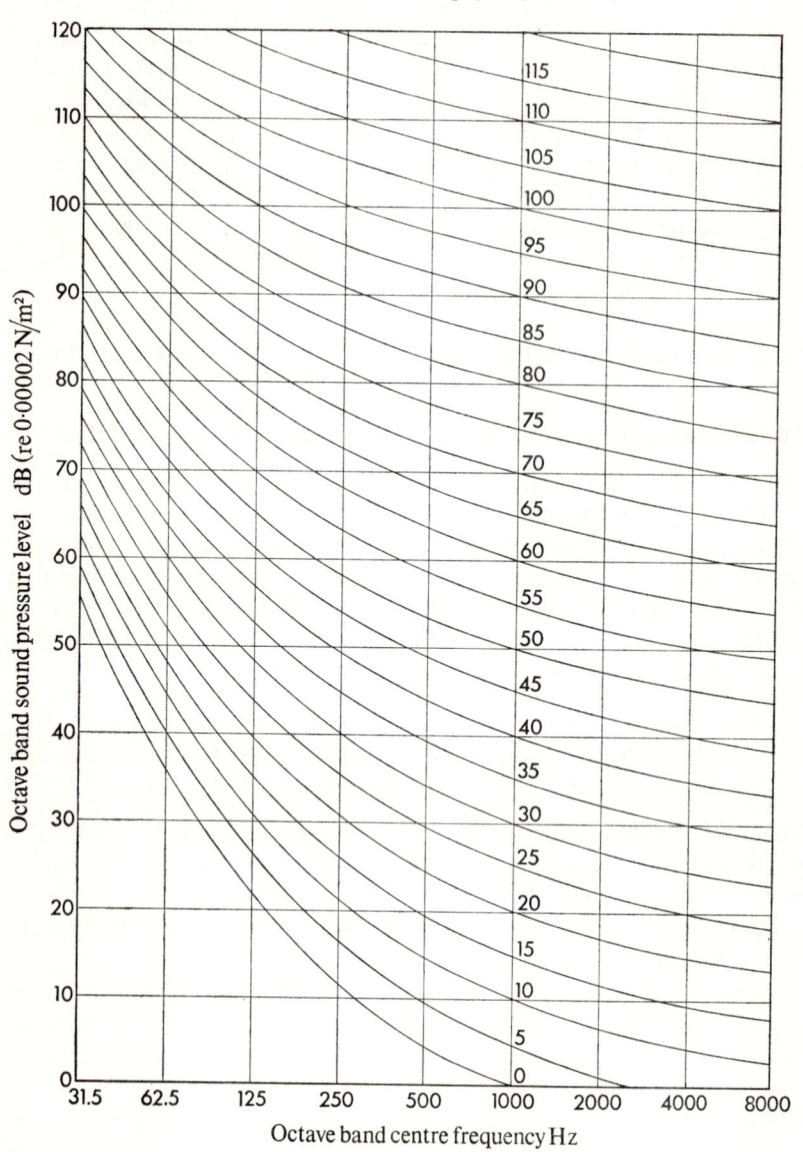

The Working Environment 133

Acoustic acceptability

Noise rating (NR) curves which show sound pressure levels in decibels (dB) at various frequencies are used to indicate suitable acoustic environments for various functions. Suggested NR curves are: conference rooms: NR 25; private offices: NR 40; offices: NR 45; workshop offices: NR 55.

Speech interference levels (SIL)

It is possible to predict the intelligibility of speech under certain noise conditions—and whether against a given amount of noise speech will be understandable at a given distance with a normal, raised, very loud or shouted voice.

Information about speech interference levels can be useful in assessing the ease or difficulty of communication under various conditions of noise.

Further reading

Environment and Human Efficiency, E. C. Poulton, Charles C. Thomas, USA (1970).
Human Engineering Guide for Equipment Designers, Wesley E. Woodson, University of California Press, Los Angeles and Cambridge University Press, London (1966).
Ergonomics, K. F. H. Murrell, Chapman and Hall, London (1969).
Occupational Health Practice, R. S. F. Schilling (ed.), Butterworth, London (1973).

CHAPTER 13

Mental Health and Motivation at Work

Work and mental health

For every case of an industrial disease caused by exposure to physical, chemical or biological hazards at work, many more are caused by the psycho-social or emotional hazards found in virtually every occupation. The subject of mental health at work has only recently become fashionable and one is tempted to suspect that this concern is caused in part, at least, by a recognition that modern industrial society has failed to meet the aspirations of an increasingly educated population, and the growing realisation that material benefits alone are insufficient.

The extensive literature on this subject seems to suffer from an alarming degree of imprecision which succeeds more to confuse than to clarify. Whilst most seem to agree on the importance of this subject, few use the same words with the same meaning, and we are very conscious that this chapter runs the risk of adding to the difficulties. Man's first needs are for food and shelter and the other basic essentials of life, but when these are satisfied, he discovers other needs such as self-fulfilment, esteem and a feeling that he can improve the quality of life for himself and his family. The concern with mental health seems to have become prominent as an indirect result of improvements in the physical environment and the standards of living in modern society. Although many of the political predictions of Marx have not been fulfilled, there is an increasing recognition that alienation of workers from the industrial system has indeed become a real issue, and one that is growing in importance.

The concept of health is a great deal more difficult to define than that of illness or injury. This is particularly true in the case of mental health. The definition of 'health' adopted by the World Health Organisation expresses the high ideals of the period after the Second World War when the organisation was established. It defines health as 'a state of complete physical, mental and social well-being and not merely the absence of disease and infirmity'. A moment's self-analysis will show that few of us are that healthy for long periods of time. Mental health is thus not merely the absence of psychiatric or emotional disorders but a more positive state of empathy with life and one's social environment, both at work and at home.

Work of some sort, though not necessarily paid employment, is important for the mental well-being of any person. It can provide a sense of purpose to life and a challenge to enable him or her to achieve ambitions and interests as well as independence. Those fortunate enough to have jobs that they enjoy, that provide opportunities for self-expression and, at least occasionally, a sense of achievement, are the fortunate few. One of the major problems of industry today is that there are many who do not get this sort of satisfaction from their jobs. For them, the hours at work are only a means of obtaining the money which allows them to obtain their satisfaction elsewhere, such as the home or garden, or from hobbies or voluntary work. A few even use quite a lot of time, energy and ingenuity in avoiding work and live on social security payments.

For most people, the requirements of a satisfying job, after ensuring adequate financial reward and safe conditions of work, can be listed:

1. It should offer some element of challenge and allow the person some freedom to take decisions.
2. It should be so arranged that he may get the support and assistance of colleagues and develop a sense of community with them.
3. It should be seen to have a desirable end product, and the job itself should be recognised as of importance to the organisation as a whole.
4. There should be some recognised status with chances of future improvements.

Dissatisfaction and frustration at work breeds anger, jealousy and the sense of alienation that are major causes of much of the industrial unrest which has become so common in countries all over the world. This is a legacy of the industrial revolution which has yet to be solved. It is no accident that much of the trouble seems to stem from factories where the production process has been so 'well' organised that people are employed almost as automata with little or no freedom to alter their manner or rate of work even though rates of pay may be relatively high. The attempts by some organisations such as Volvo or Philips to replace production line assembly work by teams who can arrange between themselves how to assemble a vehicle engine or a television set, are signs that some employers are trying to reconcile the emotional and social needs of workpeople with the needs of the organisation that employs them.

Motivation

A person may change his attitudes as his work changes, just as increasing age and responsibilities change him. Most rebellious youths are transformed when saddled with a family and a mortgage. The increasing rate of change that is characteristic of modern society, however, poses a serious threat to the needs of most for security and an assured income. Productivity deals, most popular in times of industrial expansion, inevitably threaten the job security of some and change patterns of work for others. Initial fears that automation and the introduction of computers would result in massive redundancies have not been fulfilled, but the inherent interest of many jobs and the scope for individual initiative have tended to decline. It now seems inevitable that many people will have to change their jobs, and often learn new skills, several times during their working life. Few countries or organisations have yet made adequate provision for the massive retraining and resettlement effort that will be required. Such a prospect is hard enough for a healthy and well-motivated man or woman to accept; for the mentally or physically impaired, or the socially deprived, it can appear insurmountable. Traditionally, certain occupations such as education, medicine and nursing or those concerned with religious affairs, are associated with a high sense of vocation. Insofar as this is true, such people are largely self-motivated, but a strong degree of self-motivation can also be found in some members of any occupational group. It is unwise to draw any conclusions about how people regard their job without making direct enquiry.

Job satisfaction, however it be defined, is clearly to be sought but, contrary to what is sometimes said, it does not of itself correlate closely with the quality of performance or with output. There is however a closer and inverse association with absenteeism. Social status in our society is more dependent upon what people do than what they are. This is not to argue that society does not value people for themselves and their human qualities, but it seems that these are valued less than other qualities which can be found in the materialistic, ambitious and aggressive individuals who are sometimes cited as examples of successful men. Such behaviour towards commercially-orientated goals may lead to misery for the individual, for example, following take-overs or nationalisation, or for society in terms of environmental pollution and other social problems. At the other extreme, an undue tolerance of human foibles can lead to such inefficiency in a group or organisation that it no longer remains commercially viable. The best blend of attitudes towards work, society and commercial success is continually sought and seldom achieved. Conflicts of interest are inevitable and these are most obvious in the sphere of union–management relations, since the goal of each can never for long be identical.

These points are relevant in a book about health at work since the attitudes of individuals and of groups towards work have a marked effect upon the mental health of the particular community, and the amount of absence attributed to sickness. There is, for example, a well marked relationship between the fortunes of the local football team and the attendance and output in the factory over the following days. Some experienced supervisors are able to predict with remarkable accuracy the likely number of absentees and level of bonus earnings as soon as a match is over. The influence of football can also be seen at national level during the World Cup matches, and this could well apply for cricket in the West Indies and for baseball in the United States. The period of increasing tension before a serious strike can also be measured by a rise in absenteeism and absence attributed to sickness. When the strike is over, absence rates tend to become unusually low for several weeks or even months. The cynical manager explains this by stating that it demonstrates the need of the men to recoup their lost earnings. This is far from being the whole explanation for group morale amongst the men is usually higher after a strike than it was before, even if they may still be collectively hostile towards the management.

Motivation and performance of people at work have been subjected to study in many countries and there is now a great deal of literature available. We are sometimes tempted to think that there may be as many theories as there are authors. Here we must be highly selective and indicate a few points which seem to us to be most useful in the context of this book. The idea, for example, that people can be considered as physiological automata or 'hands', dies hard. There are still many who ridicule the views of behavioural scientists and think that the secret of successful management lies in the judicial application of 'carrots and sticks' together with work study. The concept of the mechanical man who responds best to incentive payments and fear of dismissal and who can work unstintingly under ideal environmental conditions, reached its peak in the years around the First World War and is often called 'scientific management'. Its leading advocate was F. W. ('Efficiency') Taylor who demonstrated that individuals could increase their output to unprecedented levels by following his instructions on detailed methods of working and being paid by results. Unfortunately, it became increasingly clear that most groups of workpeople somehow seemed unable to behave in a similar fashion for years on end. The tide turned as a result of the now celebrated researches at Western Electric's Hawthorne works. These were designed in furtherance of Taylor's principles, with groups of assembly workers being studied under various environmental conditions and methods of organisation of work and rest. The studies required that after various test periods the work be done again under pre-test conditions. To the astonishment of all, the output remained at unprecedentedly high level. The explanation of this is now generally called 'the Hawthorne effect'. It is believed that the active interest of management and the close relationships developed with the research team in which the various changes were discussed with the employees whose views were also sought, combined to produce greater benefit than any of the planned series of changes. It must now be generally accepted that any study of work performance which is done with the workers' knowledge, particularly when their co-operation and opinions are sought, can of itself affect motivation and performance. The results are not always for the better, and the effect wears off when attention is no longer focused upon them. Industrial psychologists and sociologists have now progressed from this relatively unsophisticated interpretation of the Hawthorne study, but there are many eminent and successful management consultants whose advice and proposals can be traced back, in general principles, to this work. Recent thought has developed the theory of industry as a sociotechnical system which emphasises the mutual interdependence of both the social and psychological characteristics of employees on the one hand, with the technical and commercial features of the business on the other. One cannot safely ignore either but the difficulty lies in striking the best balance.

It is now fashionable to refer to job enrichment (since 'enlargement' could prompt demands for more pay) and a leading exponent in this field is Frederick Herzberg. He has collected a good deal of evidence to support his concept of motivation. In essence, this has further developed and modified the theory of A. H. Maslow that man has a hierarchy of needs, each of which must be satisfied before the next becomes important. Herzberg describes two different groups of factors, both of which require the attention of management if it is to obtain the best out of its work people. The first group, which he calls 'hygiene factors', includes items such as supervision, pay, job security, hours of work, environmental conditions and so on. These, he claims, must be adequate to prevent negative influences affecting employee attitudes; but even if all are satisfied, they will not necessarily result in a willing and enthusiastic workforce. This is only obtained when the second group of factors which he has termed 'motivators' are also brought in, but they do not work unless the hygiene factors are already satisfactory. These motivators include achievement and its recognition, the nature of the work itself, responsibility and the possibility of development and of advancement. The ranking order of importance amongst these factors varies from one person to the next, not only because the extent to which different organisations meet these needs, but also because people themselves differ.

The way in which work is organised and supervised, and the extent to which the capacities of the individual are utilised or stultified, will affect not only performance but also the mental health of the organisation, the department and of the individual person. One reason why scientific management was so popular with managers was that it allowed them to plan the work and manage the staff as if they were all alike. There are many people who still prefer to think on these lines today. It could, however, be argued with considerable force that many of the biggest problems arise in the management of people because they *are* all different. Standard procedure, precise rules and jobs defined to the last detail suit some but serve to frustrate others who prefer to be allowed flexibility to adapt to the needs of the moment. These individual characteristics, together with other traits, make up what we term personality. Considerable efforts have gone into the definition and measurement of human personality for many years with very limited practical success. The ability to select people of the right personality for the type of job available is still very limited. Aptitude testing for certain jobs is now quite reliable, but personality testing is certainly not. Although many organisations, notably in the United States, may make use of such tests, there are many others who have tried and discarded them.

The weaknesses of interviews are well known: the sum of repeatable information not available by other means is probably that the interviewer does or does not like the person being interviewed. Personality testing is really not much better. The almost infinite variability of human temperament and traits makes unsuitable job placement inevitable from time to time, and this is yet another cause of stress and mental ill health either for the person concerned or, just as frequently, for his colleagues and subordinates.

Stress at work

There is considerable confusion in the meanings given to the word 'stress'. To the physicist or engineer it is the force applied to a material which produces 'strain'. Some also use it in this way to describe the pressures of society which produce strain in individuals. Others, however, use the word 'stress' to describe the effects produced upon those subjected to the tensions in the psycho-social environment. This semantic confusion has succeeded in making an imprecise subject even more difficult to comprehend. Much of this confusion dates from the publications of the Canadian, Hans Selye, who ascribed many of what are now termed psychosomatic disorders to the tensions and crises of life and used the word stress to describe the consequences rather than the causes. Whatever its precise meaning, however, it describes a situation of conflict between a person and his environment.

Stress is a phenomenon which everyone experiences at one time or another. When a person feels stressed, he is under stress. What is stressing for one person may not be stressing for another; the causes of stress are, therefore, a highly individual matter. Some people may find that work which has no real beginning or end and no clear sequence or end-product may for them be totally unsatisfying and even stress-producing. Others may enjoy the challenge and lack of monotony which such a job offers and become bored, fed up and stressed in routine type work where clear goals in terms of work output or production can easily be demonstrated. Stress at work is most often related to personal relationships: with 'the boss', with colleagues or with other groups of people encountered at work. And stress may occur as easily in the manager as in the office junior; no group or person is exempt. Most people have an Achilles heel somewhere in their make-up and the particular form of stress which discovers this weakness may be found at work. To summarise, stress is an individual phenomenon, is subjective in nature and can occur in anyone who feels that he or she is under pressure.

Most people now accept that a certain amount of stress or emotional tension is necessary and indeed desirable. Difficulties arise however because it has not been possible to measure stress in quantifiable terms, and also because it is quite clear that the quantum of stress that may suit one person can be too great for another to support. The consequences of undue stress can manifest themselves in disturbances of a mental or a physical nature, or most often in a combination of the two. There are many ways that emotional tension can produce physical dysfunction; everyone has experience of this from childhood. Diarrhoea, headache, palpitations or shortness of breath are commonly found in people going through a crisis, whether this be an interview for an important job or an examination, or waiting to go into action during war. The theory underlying the concept of psychomatic disease is that persistent states of stress or tension can ultimately produce physical diseases such as duodenal ulcers, colitis, asthma and the like. Unfortunately, doctors disagree on just what conditions should be included in this list. There are those who refuse to accept that stress can be the sole cause of any of them, even though they may agree that it can aggravate conditions caused by other mechanisms.

The mystery of duodenal ulceration provides a most salutary example of the problem. The condition increased in prevalence until the early 1950s, but in the past fifteen years or so has become much less common. At one time it was widely believed to be the result of 'the pace of modern life' and emotional stress. President Truman is said to have described a Secretary of State as a 'two ulcer man in a three ulcer job'. Today, however, even if the pace of life is certainly no less, ulcers are becoming less frequent.

The effects of undue stress at work may become apparent in several different ways—from absences, injuries, poor work, irritability and increased wastage, to collective action including strikes and other industrial disputes. Depending on the manner of its presentation the problem may first come to the notice of line or personnel management or to the occupational health doctor or nurse. Problems can arise from frustration when the individual feels he is prevented from working as well, or as effectively, as he feels he is capable of doing. It can also come from over-promotion. The 'Peter Principle', which states that people are promoted until their inability to do a job becomes clear to all, may be applied, unfortunately, to many organisations. The tragedy is that over-promotion often causes a breakdown in physical or mental health. The idea that organisations too can be healthy or unhealthy has gained some acceptance and is really an extension of the military adage that the quality of a unit and its morale can be assessed by the size of the daily sick parade.

What can be done about stress illnesses when they are recognised? First, the stress or stresses can be removed from the individual or the individual can be removed from the stresses. These solutions are not always possible or even desirable and so a second line of action has most often to be taken. This is to help the individual to modify his reaction to the stress. An example may make this clear. Suppose that three people are put into adjoining and identical prison cells. The first person bangs his head on the bars, screams, scrapes his nails on the walls and finally collapses exhausted in the corner. The second person sits down in apathy and despair in the corner of the cell and weeps. The third person inspects the bars and the lock carefully, tests the bars, the lock and the walls, and concludes that physical violence directed at these will merely hurt him. He then asks when he will be allowed to exercise and when he can borrow books from the prison library. From these differing reactions to an identical situation, and there could be many other reactions, it would be possible to say something about how each person would probably survive the ordeal of imprisonment or would react to similar stresses. It may even be that the people reacted in the way they did without really thinking through the situation: the reaction was more one of feeling than of thinking. And the particular way that they reacted was conditioned more by their previous reactions and by the kind of people that they are than by any type of cold intellectual judgement. The first two people reacted in ways which stressed them. Perhaps they could be led to see that this was not the only possible reaction—there is a choice. Perhaps also, they could try to appreciate many possible reactions and go on to pick the one which would be the most useful to them, which served their ends and which did not stress them. Thus it may be possible to demonstrate to an individual who feels that he is stressed that he contributes largely to the stress which he feels. He may also be led to appreciate that by modifying his reactions he could alter the total stress situation for the better without necessarily affecting the circumstances external to him.

Lastly, sedatives and tranquillisers can be used. These will dampen down the stress reaction but will do nothing to remove the causes or to change the reactions of the individual. Tranquillisers thus have no curative effect; they are merely palliative. As such they have limited usefulness and merely allow time for the cause to be dealt with properly. However, if tranquillisers are given without grappling with the causes of the stress and without trying to reveal and understand, and perhaps to modify the reactions of the individual, then their prescription is an evasion of responsibility for the definitive treatment of the problem. Although doctors and nurses in industry are concerned with the mental health of the enterprise as well as of individuals, they are not, of course, responsible for its organisation nor for its management policy. They can, however, help in making a diagnosis and by advising managers when particular areas which seem to be causing stress come to their attention.

Further reading

Psychology at Work, Peter B. Warr, Penguin, London (1971).

Work and the Nature of Man, F. Herzberg, Staples Press, London (1968).

The Social Psychology of Industry, J. A. C. Brown, Penguin, London (1954).

On the Quality of Working Life, N. A. B. Wilson, HMSO, London (1973).

Problems of an Industrial Society, William A. Faunce, McGraw-Hill, New York (1968).

CHAPTER 14

Some Specific Occupational Health Problems

In this chapter we have attempted to skim lightly over the surface of such health problems related to occupation, but there will of necessity be many omissions. Perhaps the reader's best approach is to regard the problems discussed as examples, and to relate them *first* to the background of the book and *second* to his own experiences and problems. We have tried to select large groups and common problems. The gaps remaining are, however, large and would require a completely different approach were we to attempt a more comprehensive coverage.

Occupation, social class and mortality

One way to assess any relationship between work and health is by the study of mortality rates. Statistical tables of mortality are produced from time to time in many countries, and some provide a breakdown between occupations. In England and Wales, for example, such tables have been a feature of the decennial censuses for over a century. The information was used to good purpose by the great social reformers of the late Victorian period, and they are used still to demonstrate the effects of social class and work upon the expectation of life. Mortality studies are only practicable in the largest organisations and even with them it can be very difficult to trace people who have left unless they are pensioners. National figures have the advantage on this score because of their size, but they suffer to some extent from imprecise occupational descriptions.

The Registrar-General's tables of occupational mortality relate to the five years around each census and the latest to be published are for 1959-63. The numerator for each rate is obtained from death certificates on which the last full-time occupation should be recorded. The denominator, or population at risk, is obtained from the census returns. Although the sources of information differ, this matters little except for one important reservation. Some men in arduous jobs become unfit to continue at their work and change to less demanding occupations in the last few years of their working life. Such changes of job usually involve a lowering of wages and it has been shown that men with slowly progressive conditions, such as chronic bronchitis and arthritis, tend to get jobs lower in the social class scale before they finally retire or die. They and their relatives, however, still think of themselves as really belonging to their main occupational group. Thus an underground coal miner may be transferred to a surface job or become an unskilled worker in another industry. At the census he will fill in his new job but after he has died his relatives may tell the registrar that he was an underground miner. This can result in a spurious inflation of the death rate for underground miners and a lowering of the rate for his final occupation. Special studies have shown that this does happen sufficiently often to account for part, but not necessarily all, of the apparent excessive death rate for this occupation and for some others, such as the armed services, in peace time.

Death rates are also calculated for five broad socio-economic classes and people are allocated to these classes by their occupation—in the case of married women by their husband's occupation. Social class I includes professional and the higher managerial occupations, class III the skilled craftsmen and class V the unskilled labouring jobs. Classes II and IV include these jobs in intermediate positions. Before being able to make a reasonable comparison of death rates in different social classes or specific occupations, allowance must be made for differences in age distribution. The most convenient way to do this is to calculate an index called the *Standardised Mortality Ratio* (SMR). This is the ratio, expressed as a percentage, of the deaths observed in the group studied to the number expected if the mortality rates by age for all men, or women, are applied. The SMRs for men and for married and single women in the five social classes for the years 1959–63 were as follows:

FIGURE 8. *Standardised Mortality Ratios*

Class	Men	Married women	Single women
I	76	77	83
II	81	83	88
III	100	102	90
IV	103	105	108
V	143	141	121
All classes	100	100	100

Some Specific Occupational Health Problems

It can be seen that people in the unskilled class had nearly half as many more deaths, and in class I only three-quarters as many deaths as the average. Even after making some allowance for the downward social mobility of the chronic sick, the differences are still considerable. This disparity between class I and V has actually increased over the past thirty years. Even though death rates overall have shown some improvement, the *relative* position of unskilled workers is worse. The overall mortality of single women is greater than that of married women between the ages of twenty-five and sixty-four, but the gradient between the social classes is not so pronounced.

The following table shows a selection of twelve occupations with SMRs well above the national average and twelve well below. The interested reader can seek further information from the relevant national statistics. Social class exerts such a strong influence upon mortality through standards of nutrition, housing, education and so on, that the SMR of men alone should not be taken as evidence of any purely occupational risk. This can be allowed for by relating the SMR of men to that of their wives, using the latter as a control for non-occupational factors.

FIGURE 9: *Standardised Mortality Ratios for Men and their Wives in Selected Occupations*

Source: Tables of Mortality, England and Wales, 1959–1963

ABOVE AVERAGE			BELOW AVERAGE		
OCCUPATION	MEN	WIVES	OCCUPATION	MEN	WIVES
Armed forces	218	135	Teachers	60	66
Coal mine: face workers	180	190	Clergy, ministers of religion	62	61
Labourers and unskilled workers	170	171	Dental practitioners	64	91
Actors, musicians, stage managers	148	107	Managers	65	65
Publicans, innkeepers	147	130	Personnel managers	67	63
Fishermen	144	113	Judges, barristers, solicitors	76	72
Construction engineers and riggers	142	120	Civil service executive officers	79	79
Stevedores, dock labourers	136	146	Draughtsmen	79	68
Garage proprietors	134	119	Printing press operators	84	90
Tailors, dressmakers	126	135	Medical practitioners	89	73
Shoemakers and repairers	125	115	Workers in rubber	89	97
Coal mine: workers above ground	124	126	Painters, sculptors, artists	92	93

The ratios for wives usually approximate well with those for men, but the marked excess amongst men in the armed forces may be due, in part, to the tendency for retired servicemen to consider themselves, and be reported as being, servicemen when they die and in part due to the higher rate of cancers, heart disease and accidental death reported amongst them. With actors, musicians and stage managers there is a very high suicide rate. Fishermen, however, are known now to have a real death rate a great deal higher than the figure of 144 would suggest, because those who die at sea are not recorded by the Registrar General. The true figure, therefore, should be 177. In particular, their standardised ratio due to accidents at work is no less than a horrifying 1726, which is more than three times the ratio experienced by face workers in coal mines and *seventeen times* greater than that for all men. Here, indeed, is a hazardous occupation where rates for high blood pressure and cancers of the lung and stomach are about twice those for all men. The high ratio for both publicans and their wives could also indicate an occupational hazard since the work is usually shared. They both have seven times the expected rate of death from cirrhosis of the liver, whilst the husbands have also unduly high rates for diseases of the heart and circulation.

Incapacity to work, arising out of or attributed to work

In the section above, occupational *mortality*, the relationship between work and death, was discussed. This section considers occupational *morbidity*, the relationship between work and illness.

In any country on any given day there will be a number of people who are not at work. It is possible in some cases to measure the number of man-days of incapacity which are attributed to sickness, to non-industrial injuries, to injuries at work and to occupational diseases. In the United Kingdom during a period of one year covering parts of 1970–71 (the latest available at the time of writing), the insured population of men lost 261·27 million days, of which 7 per cent were attributed to work; the insured population of women lost 71·94 million days, of which 3 per cent were attributed to work. When the days of incapacity attributed to work are examined, men lost 18·65 million days, of which 14 per cent were attributed to occupational diseases, and women lost 2·48 million days, of which 11 per cent were attributed to occupational diseases. This has become a general pattern over the years (see figures 2 and 3, page 65).

Executive health

Is there any good reason for believing that the executive is the victim of a set of diseases peculiar to his occupation? Does he suffer from

FIGURE 10: *Plotting noise emission by audiometry*
Contours indicate approximate sound level dB(A)

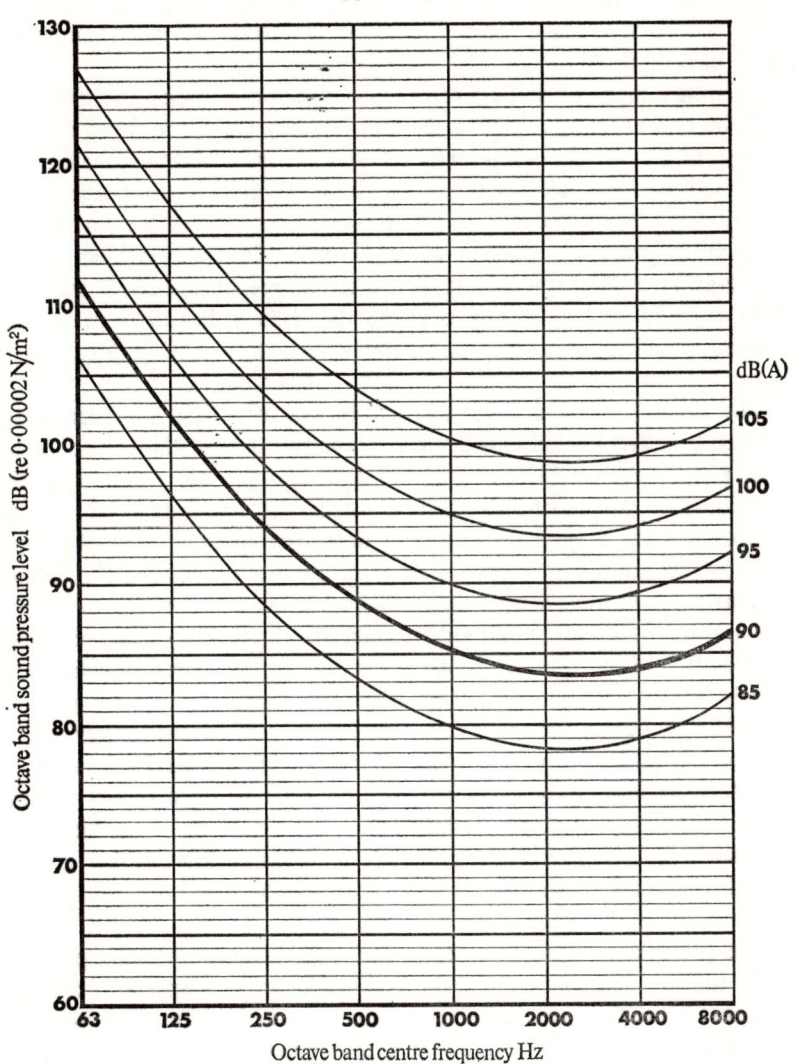

more of any disease or from any disease in a worse form than other relatively desk-bound people? The answer in both cases seems to be negative.

Studies over a five-year period of the health of executives and non-executives who worked in the same offices showed that the incidence of high blood pressure or heart disease bore no relation to the level of responsibility. It also showed, surprisingly, that among the executives there was a significantly lower amount of arteriosclerosis than would have been expected.

So much then for differences to be expected between the executive and the non-executive and for the contributory factors to heart diseases in executives of gin with large lunches, obesity, flabby toneless unexercised muscles and excessive smoking. While all these things can contribute to the incidence of heart disease, executives are not noticeably more prone than other office workers. Many executives are of above-average intelligence and are used to responsibility. Should not such a group be better able to assess the risks in health from varying causes, including over-indulgence and a sedentary way of life, and to act accordingly in their own interest?

The rôle of stress and its amount in executive life has been exaggerated greatly in the minds both of executives and of the public. Stress is a phenomenon which is largely dependent on the reaction of the individual to his environment. The blame for stress may be laid on environmental causes, for example, deadlines to meet, unsatisfactory performance of plant or people or unprofitable business. But a degree of tension is normal with acts or feelings which we believe to be important. Also, different individuals can be 'loaded' to quite different degrees before the signs of protest or strain appear. Stress, therefore, is largely a personal and subjective matter and may occur as readily in the office boy as in the managing director. It is generated largely within the person and not by age, occupation, or job demands. Tension is therefore harmful only to those who, on account of personal factors, react abnormally to stress. It has been shown that the so-called stress diseases show no preferences for high, medium or low income groups, and that executives show a similar incidence of stress diseases to that of similar but non-executive office workers.

The problem of executive health therefore equates with the problem of health in sedentary occupations. There are no special executive diseases: 'stress' as an executive disease is part of executive mythology. Would some executives feel guilty if they couldn't plead 'stress'? How far does personal inadequacy contribute to stress; and how easy is this to face up to?

The health recipe for the sedentary worker (that is, executive) is well-known. Common sense, exercise and moderation are the chief ingredients. Interests outside of work have a protective bearing and a perspective-giving function to the total business of living.

The health of drivers

Road injuries are now one of the commonest causes of death among young adults, and with coronary heart disease and cancer of the lung, form the modern epidemic killers. Studies in several countries have shown that disease of drivers, with the major exception of alcohol, is of relatively little numerical importance as a cause of road traffic 'accidents'. This may be due in part not only to the fact that virtually every country has laws requiring minimum standards of vision and health for holders of driving licences, and restricts their issue to people with certain conditions such as epilepsy, but also to the ability of most drivers who are taken ill to stop before they lose control.

Laws on driving and health usually distinguish between licences for private and for commercial driving. This is of the greatest importance, not only because of the size of the vehicle, but also because the professional driver may be at the wheel for eight hours or more a day, and he is usually driving to a schedule. A man who has recovered reasonably well from a heart attack may be allowed to drive his own car by his doctor, but it would be foolhardy in the extreme to allow him to drive a large passenger bus or a juggernaut lorry for forty hours a week. As always, the most difficult problems are found in the 'grey areas' such as driving a light delivery van for a few hours a week. In coming to a decision a doctor should take into account factors such as the type of vehicle and the effort required to drive it, the hours spent at the wheel and the traffic conditions, as well as the more strictly medical findings. The employer, however, must also take note of the legal position since he may well be held responsible when one of his vehicles is involved in an accident. Employers are usually required to take all reasonable steps to ensure that their drivers are fit to do their job safely.

Preplacement medical examinations for drivers are therefore required by many organisations. The factory doctor then becomes responsible for advising managers on the fitness of staff to drive on duty, and this can sometimes prove onerous. So much of medicine is neither black nor white and it is thus a matter of balancing risks and probabilities. There are occasions when risks to passengers or other road users must outweigh a reluctance to deprive a man of his normal job. The professional associations in several countries issue guidelines to their doctors and in Britain this has also been done in an authoritative booklet by the Medical Commission on Accident Prevention.

There are also health problems directly caused by driving. Fatigue and monotony leading to inattention or falling asleep at the wheel certainly cause accidents, even though they may seldom appear in official statistics. These can be aggravated by the all too prevalent lack of good ergonomic design of seats and controls. Manufacturers generally have paid little attention to this matter, but if they were ever held liable for the many cases of backache that arise in drivers the situation could alter dramatically. Inadequate ventilation and excessive noise also play important roles in the generation of fatigue; both are capable of solution, but only at a cost. Petrol engines pollute the environment with carbon monoxide and lead, but diesel engines, provided they are properly adjusted, create little pollution. For indoor operation in warehouses and so on, it is necessary to use electrical power for vehicles if air pollution is to be completely avoided. Finally, the use of seat belts by all vehicle occupants is now so well-known as an effective way of saving lives and reducing the severity of injuries that we feel no need to press the case further.

Food handlers

Most outbreaks of food poisoning affecting several people at the same time can be traced to faulty hygiene either in the production, handling and storage of food, or to the food handler himself. The term 'food poisoning' normally describes an acute digestive illness marked by vomiting or diarrhoea (or both), sometimes coupled with a more generalised illness and fever. Contamination of food by bacteria is the usual cause. Bacteria may produce illness either because they form toxins which cause illness within a few hours or, less frequently, through an infection by the bacteria themselves. Most of the severe outbreaks affecting many people with dramatic suddenness are due to the consumption of food or milk-based sauces or creams in which bacteria have already produced toxins.

Food handlers who suffer from an infection, for example, tonsillitis, a septic finger or gastro-enteritis, can easily transfer some of their bacteria to food, and thus to anyone who consumes it. One of the most explosive outbreaks of diarrhoea and severe vomiting which the authors have seen was caused by a cook who had a small and only mildly septic-looking cut on his finger. Another was a very large outbreak of streptococcal sore throat which put a major operational airfield out of action for a few days. It was all caused by one man with a sore throat who infected the milk. Tepid food, particularly custards, creams and warmed-up meat or poultry provide the perfect culture medium in which bacteria can grow at record speeds. Many of the intestinal infections are kept going by following the merry-go-round of hand-mouth-anus-hand-mouth-anus. Infection by this means could be prevented entirely if the basic rules of hygiene, especially the rule about washing after visits to the lavatory were followed.

The pre-employment selection of food handlers necessitates a careful enquiry, often with a written questionnaire, into previous illnesses, and a health screening check which can be done very well by a trained nurse. If any abnormality is found, the person may need to be referred to a doctor for further assessment. All who are passed as fit for employment as food handlers must also be educated or re-educated in the basic rules of hygiene, which must be scrupulously observed. The need to report intestinal or septic illnesses must be stressed to reduce the risk of affecting those who eat the food. Wise employers will make this reporting as easy as possible, and should ensure that unreasonable financial loss does not follow if a food handler is suspended from duty to protect others. Bacteriological investigations, for example stool cultures, may sometimes be required of food handlers as a necessary preliminary to employment in a food factory or elsewhere. Certain unhygienic personal habits such as nail-biting may be a cause for rejection. Personal hygiene of a high order can only be expected if suitable provisions are made including easily accessible washing and lavatory facilities of a good standard, together with an adequate supply of soap, nail brushes and paper towels. Clean overalls are also essential and there must be somewhere suitable for changing into them. Above all, the catering supervisor must be adequately trained and encouraged to call for professional advice whenever necessary.

Manual work and back pain

Man's erect posture makes him particularly liable to various painful conditions affecting the spine and its muscles. All studies of incapacity in industry have shown the high incidence of 'back trouble' as a cause of time off work, permanent limitation of duties, and of premature retirement. Although most cases occur among those who have to move heavy weights at work, sedentary office workers and vehicle drivers may also suffer from back pain. Fashions in diagnosis, and the precision in which it is made, vary; yesterday's 'lumbago' becomes today's 'slipped disc', while 'sacro-iliac strain' enjoyed considerable vogue a few years ago. The bones of the vertebral column are separated from each other by fibrous discs which have a soft centre, not dissimilar to the structure of a golf ball. Although most cases of back pain are due to sprains or strains of muscles or joints, a few are due to slight movement of these discs which may, on occasion, press on a nerve and cause pain to be felt down the leg (sciatica), in the neck, or down the arm.

Many of the cases of back pain seen by doctors present problems in diagnosis. A few may be due to trouble elsewhere, such as in the kidneys or in gynaecological disorders; others may show minor developmental bone abnormalities on X-ray which seem to be associated with an increased prevalence of backache; still others may show nothing at all. Except in acute episodes when help is sought at once, it can be difficult or even impossible to decide whether the condition was caused by an injury at work or perhaps by unaccustomed and enthusiastic gardening at the weekend. Backache, like the headache, is a common cause of absence from work attributed to sickness since, as sufferers will know, it leaves a fairly wide margin for those who do not like their job to indulge in a week or two more away in which to enjoy their incapacity for work. It seems, however, that doctors who have interested themselves in the early and active management of back pain achieve better and more rapid results than their *laissez-faire* colleagues. There is no doubt that manipulation can have excellent results in suitable cases. What, however, can be done to prevent this troublesome condition?

In a number of work situations, back pain can arise from faulty posture, from bad techniques of lifting or from working in cramped positions. Severe pain can come on quite suddenly, however, in apparently innocuous circumstances such as when an unsuspected sneeze or cough occurs whilst bending to pick up something as light as a piece of paper. Preventing back injuries and backache at work should firstly be considered as an ergonomic problem. Better design can reduce or even eliminate the hazard. In a chemical plant a few years ago, several of the operators developed backache. In many of them it was caused by the opening or shutting of manually-operated wheel valves which were positioned in such a way that operators had to bend over and twist to turn them. At the design stage of the plant the valves had been marked by the draughtsmen with an 'X' on the plans, without thought to their exact positioning. The cost of resiting a valve is considerable. Forethought at the design stage would have cost nothing and could have saved subsequent lost time from backaches. Another way in which some back injuries can be prevented is to avoid heavy lifting by introducing mechanical handling devices, conveyors or hoists. Where these are brought into use after the men have already become used to lifting the objects concerned by hand, there may be a problem in re-education, since many try out the hoist once or twice, and then revert to using their own muscles because, they say, it is less trouble!

Finally, whenever manual handling is necessary, instruction in the techniques of lifting correctly can make a significant contribution to the prevention of back strain. The worst possible position in which to put strain on the back, whether by lifting, pushing or pulling, is a bent position combined with rotation to one or other side. In this situation the vertebral column is structually unstable and relies upon muscular support. A straight back, on the other hand, results in the natural locking of the vertebrae into each other to produce stability. The weight should always be carried as close to the body's centre of gravity as possible and thus 'a straight back and bent knees for all lifting' is useful basic advice.

The value of training in correct methods of manual handling is now well-recognised and courses of instruction for this purpose are now being used in many companies. These techniques are often known as *kinetic handling*. As with the men who ignore lifting gear, the problem is to ensure that, once taught, the new techniques are used in daily life. Few people seem to take easily to using their legs to save their backs—until they themselves have suffered backache.

Occupational eye hazards

Metal workers, welders and anyone who uses chisels, grinding wheels, milling machines, high speed drills, boring equipment or chipping hammers are liable to get foreign bodies in their eyes. A few of these foreign bodies may be low speed ones, but most will be high speed fragments which will either become embedded in the cornea (the clear window at the front of the eye) or, if momentum is sufficient, will penetrate and enter the eyeball via the cornea or the sclera (the tough white outside covering of the eyeball). Many eyes are lost every year and many hospital admissions and attendances are due to foreign bodies on or in the eye. The great majority could be easily prevented by the use of suitable eye protection.

Chemicals which may damage the eyes are found in many work places. Acids are dangerous but alkalis are worse. By the time the acid has been washed out of the eye, the maximum damage has occurred and there is no persistent effect. Alkalis, however, attach themselves to the protein of the cornea and, when the alkali is washed out of the eye, some still remains. This can cause further damage. Most chemical hazards should be capable of identification; therefore the provision and wearing of suitable eye protection should prevent any possible loss of vision. One example of a chemical eye burn occurred to a chemist on the opposite side of a bench from a colleague who was using acid. The acid splashed from a broken container into the chemist's eye. In spite of immediate first-aid and skilled treatment at the eye hospital his cornea became cloudy and this resulted in no useful vision in the eye. Corneal grafts were carried out, again by excellent surgeons, but in spite of many attempts, the grafts did not take or subsequently became cloudy. In thirteen years this man spent thirteen months in hospital and finished up with no useful vision in the eye. A pair of spectacles worn at the time of the splash would have saved his eye. Some people still doubt the value of routine wearing of eye protection in areas of potential hazard. Some few people would even argue that the cost of equipping everyone with eye protection is too great. We do not agree.

Heat (infra-red radiation) is known to cause a particular kind of lens opacity, a cataract, in people who are exposed over many years. Furnacemen, glass-blowers and chain-makers are examples of the occupations that have suffered in this way. Suitable tinted glasses will prevent the condition.

Ultra-violet radiation can cause a 'sunburn' of the eye. This occurs in people exposed to the ultraviolet from electric arc welding ('arc eye' or 'welders flash') or who use sunlamps at home. So-called 'snow-blindness' is due to ultraviolet reflected from the snow on to the eyes. Dark glasses offer complete protection.

Lasers can cause serious burns of the eye—particularly of the retina. The prevention of laser injuries is discussed on p. 102.

Biological hazards associated with occupation

Biological hazards of occupation can arise from viruses and from various micro-organisms such as bacteria and yeasts, and from moulds, spores and fungi. Occasionally parasites can also give rise to problems.

A brief list of occupational diseases associated with biological hazards is given below. The list is, of necessity, incomplete.

1. *Psittacosis and ornithosis* are diseases which are normally found in birds but which can be transmitted to man, giving rise to a severe type of pneumonia. Parrots, pigeons and budgerigars commonly transmit the infection to man.
2. *Anthrax* may occur in anyone who is in contact with farm animals or with their fur and pelts. Bone meal is the commonest cause today.
3. *Brucellosis* (undulant fever) is a disease of cattle, goats and pigs which causes abortions. It is transmissible to man via milk or by direct contact with the sick animal, and gives rise to an illness characterised by intermittent fever, severe muscular pains and lethargy. Many veterinary surgeons show evidence of having had this infection.
4. *Leptospirosis* (Weil's disease, sewermen's disease) is a disease of rats which can be transmitted to man. It causes a severe form of jaundice, a high fever and sometimes spontaneous bleeding.
5. *Farmer's lung* is a disease which causes increasing breathlessness and is due to an allergic lung condition caused by the inhalation of fungal spores from mouldy hay or silage. Similar conditions have been described in pigeon fanciers, mushroom growers, maltsters and the washers of cheese cloth.
6. *Serum hepatitis* has led on several occasions to outbreaks of this

type of viral jaundice among hospital staff because of the increasing use of renal dialysis as the virus is sometimes found in the blood of healthy donors.
7. *Tuberculosis* has long been recognised as an occupational disease of doctors, nurses and medical laboratory technicians.
8. *Skin infections* of various kinds can be transmitted from animals to those who are in contact with the live or dead animal.

The difficulty with many occupational diseases of biological origin is to distinguish them from other naturally occurring disease and to be aware of their occupational origins. Only a careful history of the occupation concerned can verify whether there is a relationship between the two, and a correct diagnosis of the disease.

Occupational health for hospital staff

Like the shoemaker's children who go barefoot, the needs of hospital staff for occupational health services have been neglected for many years. Very large numbers of people work in hospitals: doctors, nurses, members of the para-medical professions, and the even more numerous lay staff in clerical, catering, domestic, portering and other essential duties. Between them they face many and varied occupational health hazards even though these may not be so obvious as they are in heavy industry. One reason for this neglect has been the ease by which doctors and nurses, and to a large extent other hospital workers too, have been able to enjoy what have been termed 'corridor consultations'. The apparent needs for traditional medical care have been met extensively on this informal basis. Treatment for nurses, particularly where they are being trained, has also been provided by hospitals on a more formal basis, but even here the changing views of staff make a system dominated by the nursing hierarchy less acceptable than it used to be. Preventive medicine has never received much more than token support in hospitals, where virtually all staff are practising treatment. The concept of occupational health has therefore been slow to be accepted in such an environment.

Occupational hazards however abound, from the wide range of physical, chemical and, above all, bacteriological risks to which laboratory and other staff are exposed, to the mundane, but often unrecognised, problems of asbestos conditions in engineers and boiler room men. The recent outbreaks of serum hepatitis, with several deaths, among staff working in renal dialysis units, and the more traditional risks of tuberculosis and other infectious diseases which are still very real, have focused attention on the whole problem. There are many other dangers which, although less publicised, also require recognition, assessment and control. These include those problems peculiar to the hospital environment such as access to drugs and the very stressful emotional situations that young nurses must learn to live with when patients, and children in particular, are dying. Drug addiction is unusually prevalent amongst doctors and nurses, although the tendency has been to ignore this whenever possible. The suicide rate, too, is high amongst them; in England and Wales for example the standardised mortality ratios (p. 144) are 177 for doctors and 133 for nurses. Medicine, nursing and allied professions may well be thought of as vocations for the dedicated but they also have their casualties, many of whom could be helped a great deal more than has been done in the past. Different problems can also be caused by the personality types of some of the non-professional staff who find their way into hospital employment. The rate of pay is traditionally low when compared with outside industry, and some people seem to be attracted to hospital work because they, consciously or unconsciously, seek a 'therapeutic community'. Some may be ex-patients who only feel secure within its doors. We must point out, however, that hospital staff do not invariably receive the best treatment, as is notoriously the case amongst doctors' families.

The situation in British hospitals, and in many other countries too, has recently altered a great deal. Many hospitals are now establishing occupational health services for their staff, which run parallel with other staff treatment services that they may also provide. As with some companies, a few appear to think that formal training in occupational health is irrelevant, and one finds that nominal responsibility is given to senior doctors and nurses in the few years before they retire. This is misguided in the extreme and real improvements in health are unlikely to result. Fortunately, most are now appointing full-time nurses, usually supported by part-time doctors, both having received training in this subject.

Medication, drugs and work

Many people come to work every day having taken some sort of medication or drug. Common drugs include simple pain relievers, such as soluble aspirin. Other drugs frequently taken are tranquillisers or anti-depressants, which can certainly influence work performance at a variety of tasks. Safety is an important question in relation to taking any sort of medication or drug when working, especially if the job is driving or some similar task involving public as well as personal safety. Strict rules may be necessary in these cases to limit medication.

Much of the self-medication which goes on is both unnecessary and ineffective. Really potent drugs can only be obtained on prescription, while the continued use of simple pain relievers can be harmful and may cause kidney damage.

Some Specific Occupational Health Problems

The best rule about taking medicines or tablets is to take nothing unless it has been prescribed by a doctor. This will avoid unnecessary medication and for many people will save a good deal of money.

Alcohol

Alcohol is a drug whose long-term effects are usually seen first in the area of social phenomena rather than in medical diseases. Certain occupations such as seamen, hoteliers and barmen have a disproportionately high number of alcoholics. Problem drinking can be identified variously: drinking under stress which proceeds to constant drinking alone, lunch-time sessions daily with reduced work capacity afterwards, and Monday absences or absences the morning after heavy drinking. Once the problem has been identified, serious attempts at rehabilitation should begin; if ignored, the situation will most probably deteriorate. Social effects of alcoholism include family stress and marriage-breakdown, loss of job, and the signs which can be seen in the individual when he fails to look after himself, such as untidiness, dirtiness, neglect of proper meals and so on. By this time, the alcoholic may suffer medical as well as social effects.

One of the problems in dealing with alcoholism is to remove the cover-up process by which alcoholics disguise their alcoholism both from themselves and from those around them. From a management viewpoint, alcoholics are often tolerated when they are turning in very poor-grade work because although they have been identified as 'drinking a bit', they are not seen clearly as alcoholics, the cover up this time affecting the observer, and are not dealt with either on a clear disciplinary basis or on the basis that they are alcoholics. Such people have a high tolerance of cosy chats and will frequently lie their way through years of this sort of tolerance before finally having to face the real problem: either to carry on drinking and take the consequences, or to stop completely.

Alcoholics, even when they stop drinking are still alcoholics. They appear to be unable to take alcohol as other people can, possibly for some biochemical reason. Once problem drinking has occurred, that person must always think of himself as an alcoholic. Others, if they are wise, will also see him wearing this label whether he is, at the time, drinking or not.

As with any addictive drug, the withdrawal period begins only when the person ceases to take the alcohol, so cutting down is both self-deceptive and useless from the point of view of solving the problem. Medical help can be given, and should be sought at as early a stage as possible, in fact as soon as the person is identified as an alcoholic, although many will continue the addiction in spite of the medical and social penalties. It is sad but true that most sufferers seem to have to await the time when they hit what is called 'rock bottom', having probably lost job, wife and friends, before really wanting treatment. The world-wide organisation Alcoholics Anonymous often provides the social and moral support during the difficult phase of rehabilitation.

Further reading

Registrar General's Decennial Supplement 1961, Occupational Mortality Tables, HMSO, London (1971).

Social and Economic Factors Affecting Mortality, B. Benjamin, Mouton and Co., The Hague (1965).

Medical Aspects of Fitness to Drive: a guide for medical practitioners P. A. B. Raffle (ed.), 2nd ed., The Medical Commission on Accident Prevention, London (1971).

Food-Borne Infections and Intoxications, Hans Riemann (ed.), Academic Press, New York and London (1969).

Rheumatism and Arthritis in Britain, Office of Health Economics (1973).

Appendix 1

Some journals on occupational health published in English

Although articles on various topical aspects of occupational health are sometimes published in journals for line, personnel or engineering managers, these articles are usually written for a non-specialist audience and do not include scientific references. As we hope this book has made clear, the practice of occupational health involves the expertise of several different professional groups including not only medicine and nursing, but also occupational hygiene (environmental engineering), ergonomics, safety and so on. With the interdependence of all scientific disciplines today, the fundamental research may derive from work in chemistry, physics, biology, pathology, engineering, psychology and many other subjects. Scientific papers of great importance to occupational health may therefore be found in an enormous range of publications. Most however tend to be published or referred to sooner or later in those journals which may be found on the shelves of an occupational health reference library. For those readers who are interested, listed below, in six groups according to the profession mainly concerned, are some of the more important journals published in Britain and the United States. Those which publish mainly original scientific research papers are marked with an asterisk.

Doctors

Journal of the Society of Occupational Medicine, published quarterly by J. Wright, Bristol.
**British Journal of Industrial Medicine,* published quarterly by the British Medical Association, London.
Journal of Occupational Medicine, published monthly by Mayo Publications, Downers Grove, Illinois.
**Archives of Environmental Health,* published monthly by the American Medical Association, Chicago.

Nurses

Occupational Health, published monthly by Macmillan Journals, London.
Occupational Health Nursing, published monthly by C. B. Slack, Thorofare, New Jersey.

Occupational hygienists

*The Annals of Occupational Hygiene, published quarterly by Pergamon Press, Oxford.
*American Industrial Hygiene Association Journal, published monthly by Bruce Publishing, St. Paul, Minnesota.

Ergonomists

*Ergonomics, published bi-monthly by Taylor & Francis, London.
Applied Ergonomics, published quarterly by I.P.C. Business Press, Guildford, Surrey.

Occupational psychologists

Occupational Psychology, published quarterly by the British Psychological Society, London.

Safety officers

Protection, published monthly by Alan Osborne, London.
Occupational Safety and Health, published monthly by the Royal Society for the Prevention of Accidents, London.

Appendix II

A draft form which could be used for investigating and reporting on an incident, and on any resultant injury and damage.

To be filled in by the supervisor directly in charge of men or equipment involved. Copy to department manager and safety adviser.

Incident/injury/damage investigation report

Personal injury	*Property damage*
☐ no personal injury	☐ no property damage
person injured	equipment involved
nature of injury	nature of damage
occupation	estimated cost to repair or replace
time	time
place	place

How did the incident/injury/damage occur?

Tick unsafe acts or unsafe conditions which led to incident/injury/damage

Unsafe acts

1. ☐ operating equipment without authority or failure to secure equipment or warn others.
2. ☐ operating equipment unsafely.
3. ☐ making safety devices inoperative.
4. ☐ using hands instead of equipment.
   using unsafe equipment.
   using equipment unsafely.
5. ☐ unsafe loading, placing, mixing or combining of materials.
6. ☐ taking unsafe position or posture.
7. ☐ working on dangerous or moving equipment.
8. ☐ failure to wear protective clothing or devices.
9. ☐ horseplay, distracting, practical joking, etc.
10. ☐ other unsafe act—*explain*

Unsafe conditions

1. ☐ improperly guarded equipment or materials.
2. ☐ defective equipment or materials.
3. ☐ hazardous arrangements or procedures.
4. ☐ problems of lighting.
5. ☐ problems of ventilation.
6. ☐ unsafe equipment or dress.
7. ☐ unsafe work system.
8. ☐ lack of skill, training or knowledge.
9. ☐ other unsafe conditions—*explain*.

Explain what was done unsafely? (see list of unsafe acts above).

Explain what was defective or unsafe? (see list of unsafe conditions above).

Appendix II

The incident/injury/damage was mainly caused by:

tick ☐ unsafe act(s)
one ☐ unsafe condition(s)
only ☐ unsafe act(s) and unsafe condition(s)

What action is necessary to prevent this happening again?

This is what I shall do to prevent this happening again:

Signature of supervisor directly in charge ....

Comments by department manager

Signature of department manager ....

Index

Absence, 33–42, 136
    attributed to sickness, 27, 36
    attitudes to, 40
absences, 33
    controlling, 36, 40
    causes, 38
    measuring, 37
    predicting, 40, 51, 137
acetyl choline, 91
addiction:
    to alcohol, 157
    to drugs, 156
Agricola, 13
air pressure, high & low, 111
allergic reactions, 95
alcohol, 157
alcoholics, 157
aldrin, 90
amines, aromatic, 93
ammonia, 88
aniline, 90
anthrax, 154
alpha particles, 98
arc eye, 101, 153
arsenic, 78, 85, 93, 94
arsine, 79, 85
asbestos, 73, 78, 80, 81, 93
    blue, 32, 81
    crocidolite, 81
    chrysotile, 81
    mesothelioma, 78, 81, 93
asbestosis, 78, 81, 83

atropine, 91
attitudes towards safety, 122
audiometry, 106
automation, 135

Back pain, 151–3
bagassosis, 81
barrier cream, 76
Bell, Joseph, 94
bends, 111
benzene, 7, 73, 89
benzidine, 93
beams, laser, 102, 154
beta rays, 98
bichromates, 96
biological hazards, 154–5
biological gradient, 63
    plausibility, 63
    coherence, 63
    monitoring, 32
brucellosis, 154
byssinosis, 82

Cadmium, 85
cancer:
    of bladder, 32, 92–3
    of lung, 13, 78, 92–3
    by occupations, 92
    of pleura, 78, 81, 93
    of peritoneum, 81, 93
    of skin, 76, 94, 101
carbon dioxide, 68, 87

Index

carbon bisulphide, 90
carbon monoxide, 79, 86
carbon tetrachloride, 73, 89–90
carborundum, 73
Carson, Rachel, 90
carcinogenic chemicals, 92–4
cataract, 101, 154
catering staff, 150–1
causes of absence, 38–9
certificates, medical, 34–6
chemical hazards, 78–97
chemical hazards, detection
    and measurements, 96
cholinesterase, 91
chlorine, 79, 88
chloroform, 90
chromic acid mist, 79
circadian rhythms, 58
coalworkers' pneumoconiosis,
    81
codes of practice, 8
cold:
    environments, 110
    stress, 110
    working in, 110
control of hazards, 71
confidentiality, 29
copper, 78
cotton, 82–3
courses—special training, 22
creosote, 91, 93, 101
critical event, 116

Damage, 113
DDT, 90
decibel, 103
decompression sickness, 111
defatting of skin, 95
dermatitis, 68, 84, 89, 94–5
derris, 90
design engineering, 128, 150, 152
diatomaceous earth, 81
dichlorvos, 91
dieldrin, 90

diesel engines, 150
diisocyanates, 8
disability, 43–52
    forecasts, 44
    disabled people,
        employment of, 43–52
dissatisfaction at work, 135–6
DNOC, 91
dosimeter, noise, 104
double-day shift, 54
driver's health, 45, 149, 150
drugs at work, 156–7
dust, 78, 80

Ear:
    moulds, 106
    muffs, 106
    plugs, 106
education in hazards of work, 71
enclosure, 74
endrin, 90
environments at work, 127–33
epoxy glue, 7, 96
ergonomics, 14, 23, 128–33, 152
ethics, medical, 22
evaluation of hazards, 69
executive health, 10, 31, 146, 148
eye hazards, 153–4

Farmer's lung, 83, 154
film badges, 100
first-aid, 23–4, 40
fitness for work, 26
food handlers, 150–1
frequency rates, 37, 115
fumes, 78
function of occupational health
    services, 12

Galen, 41
gamma rays, 98
gammexane, 90
gases, 79, 86
glare, 130
glass fibre, 82

glass wool, 82, 106
glass-wool for noise protection, 106
glue sniffing, 90
gradient, biological, 63

Haematite, 82, 93
halo-ethers, 93
halogenated solvents, 89
hatters' shakes, 84
Hawthorne effect, 138
hazards:
   biological, 154–5
   chemical, 78–97
   eye, 153–4
   of high air pressures, 111
   of high temperatures, 109
   of low pressures, 111
   of low temperatures, 110
   occupational, 66
   toxic, 67
health:
   declaration, 27
   definition of, 134
   of drivers, 149–150
   of executives, 10, 31, 146, 148
   interviews, 28
   of managers, 10
   responsibility for, 10
   screening, 28
   services at work, 18
   and shift work, 55–8
hearing conservation programme, 106
hepatitis, serum, 154
herbicides, 91
Herzberg, F., 138
high temperatures, 109
Hippocrates, 13
Holmes, Sherlock, 94
hospital staff, 155–6
housekeeping, 76
humidity, 109, 131
hydrogen sulphide, 79, 87

Identification of hazards, 68
illumination levels, 130–1
impairment, degrees of, 46
industrial diseases, 7
influenza immunization, 41
infra-red radiation, 101, 154
injuries at work, 64, 113–26
International Labour Organisation, 12, 94
   investigating injuries and damage, 113
incentives for safety, 125
insecticides, 90
isocyanates, 8
ionising radiations, 93, 98–101
itai-itai disease, 86

Kieselguhr, 81, 83
kinetic handling, 153

Laser beams, 102, 154
lead inorganic, 83
   organic, 84
legal problems in occupational disease, 62
legislation:
   trends of codes of practice, 8
leptospirosis, 154
light work, 26, 49
lighting, 130–1
local exhaust ventilation, 74
low air pressure, hazards of working in, 111
lumbago, 151
lung tumours, 93

Maintenance and housekeeping, 76
malathion, 91
malingering, 35
managers, 9
   responsibility for health, 10
manual work and back pain, 151–3

Index

Marx, Karl, 134
Maslow, A. H., 138
measuring safety performance, 115
measurement of absence, 37
medical:
- advisers, 21
- certificates, 26, 34–6
- ethics, 22, 29
- examinations, 25–32
- examinations, reasons for, 26
- examinations, recruitment, 26, 40, 51, 149, 151
- examinations and sickness absence, 25, 40
- examinations, techniques, 27

medication and drugs at work, 156–7
mental health and work, 134–42
mercury, 80, 84
- organic compounds, 84

mesothelioma, 78, 81
metal fume fever, 79, 86
methanol, 90
methyl chloride, 90
microwaves, 102
mineral oil, 68, 93
- wool, 82

Minnemata disease, 85
mists, 79
motivation, 41, 136–9
morbidity and occupation, 146
mortality and occupation, 143–6
multiphasic screening, 28

Naphthylamine, beta, 92–4
neutrons, 98
never sick, 39
nickel, 93, 95–6
night work, 54
nitrobenzene, 90
nitrogen, oxides of, 88
noise, 9, 103–7
- continuous, 103
- design in relation to, 105
- discontinuous, 103
- dosimeters, 104
- effects on the community, 104
- effects on work people, 103
- measuring, 104
- personal protective equipment, 106
- rating curves, 132

numbers of occupational health staff, 21

Occupation and mental health, 134–42
occupation and social class, 143–6
occupational:
- cancer, 92–4
- cancer of skin, 93–4
- deafness, 103
- disease, control of, 72
- disease, diagnosing, 62
- disease, education in, 71
- disease, legal problems, 62
- disease recognising, 62
- hazards, 62
- hazards, controlling, 71
- hazards, evaluating, 69
- hazards, identifying, 68
- health, professional training in, 22–3

health, problems at work, 7
health services, 13, 18, 20
- health service for hospital staff, 155–6
- health services, introducing, 18
- health services, personnel for, 19, 21
- hygienists, 17, 19, 23, 69
- morbidity, 146
- mortality, 143–6
- medical advisers in, 21–2
- nurses, 19, 21–2, 28
- skin cancer, 76, 94, 101
- skin diseases, 94

oil, mineral, 68, 76, 93
  shale, 93–4
organic lead compounds, 84
organic mercury compounds, 84
ornithosis, 154
organo-phosphorus
    compounds, 91
Owen, Robert, 16
oxides of nitrogen, 88
ozone, 79, 87

Parathion, 91
paraquat, 91
pension funds, 26, 51
pentachlorophenol, 91
personal hygiene, 76
personal protective devices
    against noise, 106
physical hazards, 98–112
pneumoconiosis, 80
  in coal workers, 81
polycyclic aromatic
    hydrocarbons, 93–4
Pott, Percival, 94
predicting absence, 40
pressure, high—working in, 111
pressure, low—working in, 111
pre-symptomatic diagnosis, 29–30
productivity deals, 50, 136
professional training in
    occupational health, 22–3
psittacosis, 154
psychosomatic disease, 139–40
publicity and warnings, 77
pyrethrum, 90

Radiation, 92–3, 98–102
  alpha, 98
  beta, 98
  film badges, 100
  gamma, 98
  industrial uses of ionising, 99
  infra red, 101, 154
  microwaves, 102
  permissible doses of, 99
  protection against, 100
  recording individual exposure
      to, 99
  ultraviolet, 92, 101, 154
radio waves, 102
Ramazzini, Bernadino, 13
Raynaud's phenomenon, 108
reactions, allergic, 95
recruitment medical examinations,
    26, 40, 51, 149, 151
rehabilitation, 46
rehabilitation workshops, 49
resettlement, 40, 136
responsibility for controlling
    absence, 36
responsibility for health at
    work, 10
restrictions, 44
retraining, 46, 136
rhythms, circadian, 58

Sampling for chemical hazards,
    96
safety at work, 113–26
safety problems at work, 113
safety programme, 121
screening, 25
  multiphasic, 28
  tests, 28–32
sensitisers, of skin, 96
serum hepatitis, 154
sewermen's disease, 154
Seyle, Hans, 138
shale oil, 93–4
shift systems, 54
shift work:
  and family, 59
  and health, 55–8
  social effects of, 59
  types of, 54
shifts:
  continental, 55

Index

double day, 54
metropolitan, 55
split, 54
traditional, 55
twilight, 54
sickness absence, 33–42, 137
sickness, decompression, 111
sick pay, 34
silica, free crystalline, 80
silicosis, 66, 73, 80, 83
silver mining, 13
skin infection, 155
slipped disc, 151
sniffers, 90
social class and occupation, 143–6
solvents, 89
soot, 93–4
specialist training courses, 22–3
speech interference levels, 133
standardised mortality ratio, 144
stibine, 79
stress, 139–42
substitution, 73
sulphuric acid, 79

Talc, 79–81
Taylor, F. W., 137
tar, 93
teeth, 79
temperatures, working, 131
   high, hazards of working in, 109
   low, hazards of working in, 110
TEPP, 91
tetrachloroethane, 90
tetraethyl lead, 74, 84
Thakrah, Charles Turner, 14, 16
thermal environment, 131
threshold limit values (TLVS), 70
   ceiling value, 70
   mixtures of substances, 70
   skin notation in toxic hazards, 70

TLVS, 70
toluene, 73, 89
toxic hazards, biological monitoring for, 32
toxic substances:
   absorption through the skin, 67
   inhalation of, 67
   ingestion of, 67
   routes of entry of, 67
tuberculosis, 30, 80, 155
tumour, lung, 92–3
training in occupational health, 22–3
types of shift work:
   continental, 55
   double day, 54
   metropolitan, 55
   split, 54
   traditional, 55
   twilight, 54

Ultraviolet radiation, 93, 101, 154
ultrasound, 107
unsafe acts, 118
unsafe conditions, 118

Vanadium, 68, 78
vapours, 69, 89
   from solvents, 89
ventilation:
   general, 75, 131
   in working environments, 74–5, 131
   local exhaust, 74
   veterinary surgeons, 154
vibrations, 107
vinyl chloride, 93

Warnings and publicity, 77
Weil's disease, 154
welding, 82, 88, 101, 154
wet methods, 75
white spirit, 73
wood dust, 93

working:
  environment, 127
  in cold environments, 110
  in hot environments, 109
work:
  shift, 53–61
  and mental health, 134–42
  night, 54

workshops, rehabilitation, 49
World Health Organisation, 12, 134

X-rays, 93, 98
xylene, 73, 89

Zinc, 78